Vauxhalls come to "The Port"

The story of the development of Hooton Park, Cheshire into the Vauxhall Motors Ltd., Ellesmere Port Plant.

Bill Thacker

First published in the United Kingdom, 2010,
by Stenlake Publishing Ltd.
Telephone: 01290 551122
www.stenlake.co.uk

ISBN 9781840335231

Acknowledgements

Ray Bernie for advice regarding the council's views of the development, Tony Burnip for his guidance on information sources, Jane Davies for her local history knowledge of Hooton Hall, John Morton and Hilbert Talbot for drafting the trade effluent and the Press Shop sections, Jim Madden for recollections of the equipment moves from the southern plants and inter-plant transport operations, Andrew Duerden and Julie Todd for their searches in the Vauxhall photo archives, Eddie Thomas for his keen eye for typing errors, my wife, many former Vauxhall colleagues, Chester Record Office staff and volunteers and many other people for their background knowledge, support and encouragement without which I could not have written this book.

Statements made and opinions expressed in this volume are based upon the author's personal experience and research. However, any errors of fact or omissions will gladly be corrected in any subsequent editions of the book.

Every effort has been made to contact the copyright holders of the images used in this book. The author will be pleased to hear from any person he was unable to reach and full acknowledgement will be given in future reprints.

Abbreviations used:

AEU	Amalgamated Engineering Union
BMC	British Motor Corporation (after the amalgamation of Austin and Morris)
CRO	Cheshire Record Office
EPBC or EP	Ellesmere Port Borough Council
GM	General Motors Corporation
GMOO	General Motors Overseas Operations
HF&P	Howard Fairbairn and Partners Ltd. (architects)
MAC	Management Advisory Committee (Vauxhall Management/Works liaison committee)
MoHLG	Ministry of Housing and Local Government
MD	Managing Director
NUGMW	National Union of General and Municipal Workers
T&GWU	Transport and General Workers Union
VM	Vauxhall Motors Ltd.

In April 2008 Vauxhall Motors Ltd. was renamed GMUK Ltd. and the Ellesmere Port plant is now called GM Manufacturing. However, for the sake of continuity in this book the name "Vauxhall" has been retained.

Contents

Preface

Although no accurate figure can be obtained, it is estimated that about 100,000 people have been employed by Vauxhall Motors since November 1962 when the Ellesmere Port Plant opened. Some only stayed a few days and some even gave up after a couple of hours, while a few have reached forty years service and their children are now working there. This employment has had an enormous collective impact upon the lives of the people of Chester and Wirral as there cannot be anyone living in the area who doesn't know someone who has worked at Vauxhall.

For the last twelve years I have been a volunteer [worker?] in the Cheshire Record Office. In 2008 I was given the task of cataloguing the planning applications for the Vauxhall plant which had been recently deposited by the local authority, Ellesmere Port and Neston Borough Council. On my initial overview of these records, which occupied some ten feet of shelf space, I recognised an opportunity to undertake the task of writing this brief history of the background and early development of the plant.

Hooton before Vauxhall

The story of Hooton township, Hooton Park and Hooton Hall can be traced back over 900 years to the Norman conquest. The origins of the name 'Hooton' comes from the two Saxon words, '*ho*' meaning a heel or point of land stretching into the plain or sea, and '*tun*', a farm. The name appears in the Domesday book as "Hotone"; a manor of the Wirral Hundred of Cheshire and is recorded as follows:

Richard de Vernon holds Hotone. Toki held it. There are 1 hide and two [third] *parts of 1 hide that pay geld. The land is for 3 ploughs. Therer are 4 riders, 1 villagers and 4 smallholders with 2 ploughs. At the time of King Edward in 1066 it was worth 30s. later 5s. and now* [in 1086] *16s.*[1]

Toki also held other lands in the Wirral Hundred and is described as a "freeman". Shortly afterwards the lands were granted to Adam de Aldithleigh, a follower of William the Conqueror, and then passed through various families until 1310, when William de Stanley inherited it together with the Master Forestership of Wirral.

For 500 years it was the home of the Earls of Stanley, senior line of the famous Earls of Derby. Towards the end of the 15th century the Stanleys replaced the original manor house (details of which have been lost) with a half-timbered Manor House.

This in its turn was pulled down and was replaced by a mansion in about 1778, built from local Storeton stone, designed in the Italian Palladian style for the fifth Baronet, Sir William Stanley, by the fashionable architect Samuel Wyatt.[2] The Stanley family laid out the racecourse and meetings were held there regularly in the 19th century and until the First World War.

On the 14th November 1849, with family finances strained by gambling debts, the Stanleys sold the estate by auction. The successful bidder, at 82,000 guineas, was their Liverpool banker, Richard Christopher Naylor. He then spent a further 50,000 guineas enlarging the hall with the addition of a 100 foot tower, an art gallery, and a large dining hall.

Figure 1: Hooton Hall 1489-1778.

[1] John Morris, *Domesday Book*, Phillimore, Chichester, 1978

[2] W Watts, *The Seats of the Nobility and Gentry*, John and Josiah Boydell, 1780

Figure 2: Hooton Hall 1778-1854.

The hall stood on land about 100 feet above sea level and had good views across the Mersey to the Lancashire hills, Naylor also improved the racecourse by adding a grandstand and a polo ground. He also established a heronry, a stud farm and built a church in Childer Thornton in memory of his first wife who had died in childbirth and to where he transferred both their remains from the churchyard at Eastham. By 1870 the idea of a ship canal connecting Manchester with the river Mersey was being mooted. It soon became obvious that this would pass the shoreline of Hooton Hall and Naylor began prolonged legal arguments with the Board of Trade and the MSC regarding the rights to salvage wrecks on the Mersey foreshore and his land ownership.[3] In 1864 Naylor had purchased Kelmarsh Hall in Northamptonshire for its hunting potential and in 1875 he commissioned W. Clowes and Sons to organise an auction sale of most of his art collection and the residual contents of Hooton Hall. The sale lasted for ten days in August that year and Naylor probably moved his family to Kelmarsh Hall about that time. His yacht had been moored on the river Mersey but in the late 1880s the construction of the Manchester Ship Canal would have prevented access to his mooring. Surely the access problem could have been resolved with a small ferryboat? However the estate continued to be farmed and the final race meeting was held just a few days after the start of the First World War.

Figure 3: Richard Christopher Naylor.

[3] CRO Ref: D 2903/11

Figure 5: Architect's drawing of the Tower 1853.

Figure 4: Hooton Hall 1854-1925.

The hall suffered badly during occupation by the British and American forces to such an extent that after the war it was deemed uneconomical to repair and like many large country houses at that time it was demolished in 1925.

Figure 6, lower right: Part of the 1911 25" OS map showing Hooton Hall.

Figure 7, upper right: The Tower, demolished in 1925 along with the rest of the hall.

Figure 8, below: This is one of the very few known photographs of the whole monument and there do not appear to be any references to it in local records.

This unusual monument, which stood at the rear of Hooton and survived the ravages of the demolition contractors and the occupation of the airfield through two world wars is shown here. The monument had three sides and was mounted on three stone tortoises; the inscription was in Greek, but not clear enough in the photograph for translation and may even be pseudo Greek just for decoration.

I have been given several suggestions regarding the monument's origins:

1. It was a monument to a racehorse called Macaroni which was one of six horses Richard Naylor bought from the Duke of Westminster to establish his stud farm at Hooton. A descendant of Macaroni, Ormonde won the English triple crown in 1886. Only fifteen horses have ever achieved this fete. It has been suggested that the three sides of the monument reflect this success and the tortoises perversely suggest speed.

2. It was a memorial to a daughter of one of the Stanleys who died in a riding accident.

3. It was a souvenir of somebody's grand tour of Europe, as there are similar statues in Athens.

4. It was a garden ornament given to Naylor by a member of the Grenfell family.

Figure 9 (above) shows officers of the King's Regiment (Liverpool) sitting on some steps with the monument behind them. The steps lead up from a sunken garden on the terrace and can be seen on the 1911 map (figure 6). Study of maps and the few photographs that exist suggest that it appears to have been located at the apex of the grass terrace of Hooton Hall facing north east. It does not appear in the sale inventory for the hall of 1849, nor is the terrace shown on the tithe apportionment map *circa* 1850. Therefore it probably arrived at Hooton in the latter half of the 19th century.

Similar monuments are known to exist at Lucan House, Dublin (1771), Stanmer Park, Brighton (1775), Lord Yarborough's Park at Brocklesby, Lincolnshire (1785) and at Mount Edgcumbe, Plymouth (1791). Photographs of these clearly show an urn on the top adding another two feet to the overall height. Figure 10 is typical of the others. The earliest of these (at Dublin) is made from Portland stone and is presumably the prototype for the other examples which were made from Coade artificial stone at Lambeth in London during the late 18th century. Coade stone was a mixture of 10% grog, 5-10% crushed flint, 5-10% fine sand, 10% crushed soda lime glass, and 60-70% ball clay. This was pressed into a mould and then fired in a similar way to pottery, but due to the size of the monuments and the difficulty controlling kiln temperatures at the time this must have been an extremely skilled operation. By using moulds several identical copies of an item could be made.

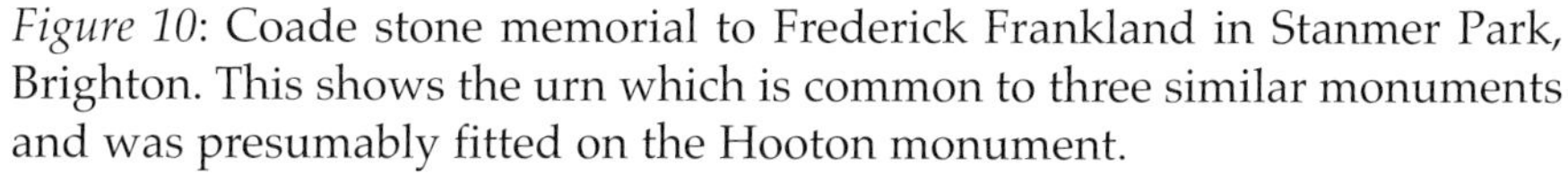

Figure 10: Coade stone memorial to Frederick Frankland in Stanmer Park, Brighton. This shows the urn which is common to three similar monuments and was presumably fitted on the Hooton monument.

Figure 11: The yard of Mrs. Coade's Stone works.

The picture of Mrs. Coade's factory in Lambeth shows some examples of her work which were often on a massive scale and show similarities in style to the Hooton monument, particularly the urn on the left of the picture. Other notable examples of Coade stone are the lions at the end of Westminster Bridge in London and the facings of Buckingham Palace.

Reviewing the four theories put forward for the monument's original purpose, the evidence of five identical monuments would seem to rule out the "souvenir from Greece" idea. Also since no daughter of the Stanley family is known to have been killed while riding, the "accident" theory can also be ruled out. Since the monument predates Naylor's interest in the turf by 60 years it rules out the "Macaroni" theory. As Naylor was an avid collector of works of art this only leaves the likelihood that he acquired it as a second-hand garden ornament, and what better place for him to put it than on his new terrace.

Where the monument went to after Vauxhall took ownership of the site remained a mystery to the author for many months. It had definitely been removed, as the ground on which it stood was excavated to provide a car park for production cars and the spoil was used for the M53 access ramps. There were rumours that it had been removed by a senior executive to his house or even shipped back to the States. An explanation was eventually found in a *Chester Chronicle Supplement* printed in January 1982 to celebrate twenty years of Vauxhall at Ellesmere Port. This reproduced an old photo of the monument and gave the following explanation of the monument's disappearance:

The monument, which once stood in the grounds of Hooton Park, is a complete mystery. When Vauxhall took over the site they were still trying to figure out what it commemorated when bulldozers moved in and crushed it to rubble. It is now buried in pieces somewhere under the factory, and the company has been unable to learn just what it was doing there.

This was surely a case of archaeological vandalism.

The First Military Use of the Site

War was declared against Germany on 4th August 1914 and Hooton Park's final race meeting was held on Friday 14th August. These two events really marked the end of the aristocratic, cultured and turbulent life that the private residents of Hooton Park had seen for nearly five centuries through successive generations of the Stanley family and then finally the Naylors.

Over the next century many different businesses would rise and flourish here and some would fail. Thousands of military and civilian personnel would learn new skills, develop talents, build careers and earn their livelihoods from the various enterprises, large and small, which came to Hooton Park subsequently. For the last 48 years Vauxhall Motors have been the main occupier of the park. This is the early part of their story.

The War Department soon requisitioned the whole estate for use as an army training ground. The hall became a headquarters, hospital, and officers' mess. Lord Derby recruited the first Pals Regiment and Hooton became the training ground for the 18th Battalion of the King's Regiment (Liverpool). They left for France and fought in the Battle of the Somme between July and November 1916.

Figure 12: Hooton Hall's main art gallery.

In Figure 12 the man on the right is sitting on his mattress and more bedding is piled up by the pillar on the left side. These are presumably new recruits as none seems to have been issued with any items of uniform.

The War Department built three double aircraft hangars which were completed in 1917. These hangars have a unique latticed timber roof construction – Belfast Trusses – with no internal supports. This method of construction was originally used in the Belfast shipyards to cover large working areas without internal supports. No. 1 Hangar was later used as a service department for Vauxhall's fleet of cars and trucks. In addition to the regular servicing work, special investigations were carried out on cars which had been returned by dealers with intractable problems.

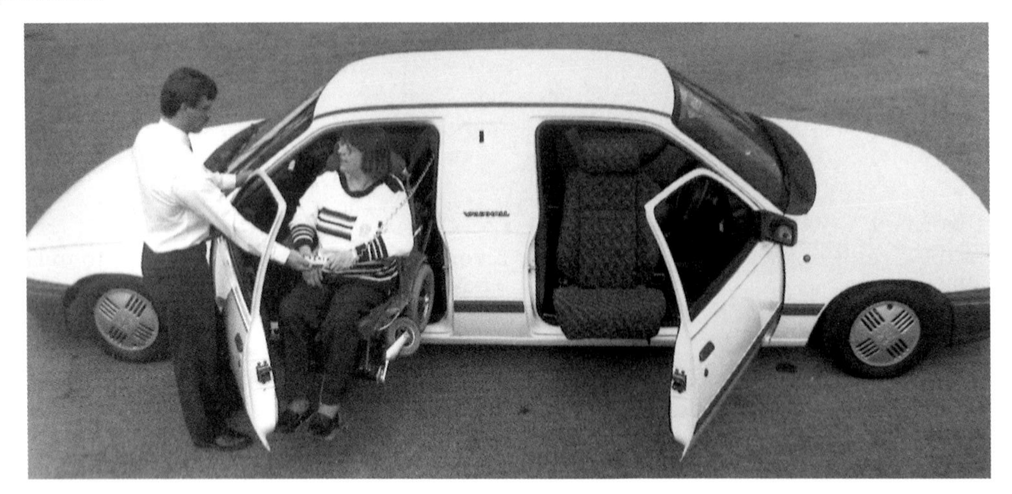

Figure 13: This double-fronted Astra was produced for assessing disabled driver's capabilities and needs. It had four driving positions each with a different type of seating and controls and could be driven from any of the four positions. Any specialist work on cars such as preparation of cars for shows and exhibitions and specialist conversions was also carried out in this department. These hangars are now classified as Listed Buildings of Special Architectural or Historic Interest and are under the care and protection of the Hooton Trust Ltd.

The Royal Flying Corps moved in and Hooton Park Airfield became No.4 Training Depot Station to form the fighter squadrons so badly needed in France. They were equipped with Sopwith Scouts, Sopwith Dolphins and Avro 504s. Some of the pilots were killed in training accidents and now lie buried in the local churchyard at nearby Eastham. A large number of American and Canadian pilots were also trained at Hooton Park. The hall was used for officers recovering from injuries.[4]

During the First World War Mr. Reginald Poole, a qualified surveyor who had previously worked for the Naylor Estates, was released from active service to return to Hooton and was charged with the task of laying out North Road to provide a link between Ellesmere Port docks area and the A41 near Eastham village.[5]

In April 1918 the Royal Flying Corps became the Royal Air Force. At the end of the war, the 37 aircraft were moved to nearby R.A.F. Sealand. R.A.F. Hooton Park was closed and the aerodrome reverted to farmland. The hangars were empty and the Hall was so damaged by military use that it was sold as a redevelopment opportunity. It was subsequently demolished although the racecourse and polo ground remained.

In the last stages of the war aircraft mechanic William Barker who had enlisted in the R.A.F. in September 1918 was posted to Hooton. Being under age for active service he served on important work in connection with aircraft repair until January 1919 when he was demobilised.[6] After many years of service at Vauxhall Motors in Luton he returned to Hooton in 1962 as tool room supervisor Bill Barker.[7]

[4] Peter Richardson, *Hooton Park a Thousand Years of History*, Hooton Airword, South Wirral, 1993 and Wikipedia (http://en.wikipedia.org/wiki/hootonpark)

[5] Interview with Reginald Poole's son, Peter, lately agent to The Trustees of the Naylor Estates, 15th May 2009

[6] *International Roll of the Great War 1914-1918*, Section V, p.19

[7] Bill Barker's retirement speech, 1965

Civil Aviation Use between the Wars

The airfield was purchased by an air enthusiast, Mr. G. A. Dawson and the land reverted to pasture but the racecourse and polo ground remained in occasional use. In the summer of 1927, Liverpool Corporation held an air pageant at Hooton as part of its civic week. This show was such a success that the Liverpool and District Aero Club was formed at Hooton and Mr. Dawson allowed the new club to use his aerodrome. In just twelve months the club became one of the most successful in the country and was the centre for aviation in the north. In 1932 an edition of *The Aeroplane* magazine a correspondent wrote that ...*Hooton is a particularly suitable place for aeronautical activities in any form... In fact I have not seen an aerodrome outside London which is so well suited for the private aviator, for an aircraft operating concern or for aircraft manufacture.*

For three years the aerodrome served as Liverpool's airport until a new airport at Speke was opened in 1933 on the north side of the river Mersey and was operated by Liverpool Corporation. Liverpool's John Lennon Airport is now on the adjacent site.

In 1934 Midland & Scottish Air Ferries offered regular services from Hooton to the Isle of Man & Belfast, Renfrew & Glasgow and Birmingham & London. The Hooton to Belfast fare was £3 single and £5 return, a considerable sum at the time. The flight took 3 hours 15 minutes, including a 10 minute stop in the Isle of Man.[8]

Dawson persuaded two R.A.F. engineering officers to resign their commissions and set up their own companies at Hooton – Nicholas Comper, who designed and built the Comper Swift, a single engined sporting monoplane and Douglas R. Pobjoy who manufactured the seven cylinder radial engines for Comper in huts alongside Hangar No 2. In 1933 Dawson ran into financial trouble.[9] The flying club subsequently moved to Speke for cheaper hangar and clubhouse facilities.

Comper removed the company from Hooton to Heston (near to the present day Heathrow Airport) in an effort to secure better production facilities, but the company folded in 1934. He died in 1939 aged 42 at Hythe in Kent as the result of a stupid prank (he was well-known for practical jokes). Apparently he threw a firework in the street, and shouted "I'm going to blow up the Town Hall, I'm an IRA man!" A passer-by took exception to this and punched him, resulting in Comper falling down and hitting his head on the kerb causing a brain haemorrhage from which he later died.

Pobjoy went to work for Short Brothers at Rochester, but was killed in a mid-air collision in 1946.[10]

Despite these setbacks Hooton was still an important aerodrome with several small commercial operators and many private owners continuing to fly from there. In 1935, Martin Hearn, an ex-pilot and ground engineer created Martin Hearn Ltd., employing a few mechanics to service the aircraft using the aerodrome. He had previously worked for Cobham's Flying Circus as a wing walker and aerial trapeze artist.

In 1936, as the war clouds gathered, 610 (County of Chester) Squadron Auxiliary Air Force was formed at Hooton Park. Most of the prospective pilots took private flying lessons in order to qualify. One person said, "Never have I seen so many Rolls-Royce cars in one spot at the same time", an indication of the typical pilot's social status. The unit was initially a bomber squadron equipped with Hawker Hind and Hart bombers and Avro Tutor trainers. In 1939 the squadron took charge of a flight of Hurricanes which was quickly replaced by Mark I Spitfires.[11]

[8] Advert in the *Ellesmere Port Pioneer*, 13th March 1934

[9] For details of Comper and his aircraft see http://nickcomper.co.uk/the-comper-aircraft-company

[10] Typescript memoirs of Ron Hosie, CRO Ref: D 6432

[11] Peter Richardson, *Hooton Park a Thousand Years of History*, Hooton Airword, South Wirral, 1993 and Wikipedia (http://en.wikipedia.org/wiki/hootonpark)

Midland & Scottish Air Ferries

LIMITED

HOOTON ∴ TELEPHONE 197

Our New Service is now in operation.

Hooton to Isle of Man & Belfast

	a.m.	p.m.
Hooton (dep.)	9.50	3.50
Isle of Man (arr.)	11.15	5.15
Isle of Man. (dep.)	11.25	5.25
	p.m.	
Belfast (arr.)	1. 5	7. 5

Renfrew & Glasgow

Hooton (dep.)	9.50	3.50
Renfrew (arr.)	12.15	6.15
Glasgow (arr.)	12.35	6.35

Birmingham - London

	a.m.	p.m.
Hooton (dep.)	9.30	3.50
Birmingham (arr.)	10.50	5. 0
London (arr.)	12. 5	6.15

Fares from Hooton

	Single.	Return.
LONDON	70/-	105/-
BIRMINGHAM	30/-	45/-
ISLE OF MAN	30/-	50/-
BELFAST	60/-	100/-
GLASGOW	75/-	110/-

★ SAFETY - COMFORT - SPEED ★

Figure 14, left: An advert from Ellesmere Port Pioneer in 1934.

Figure 15, below: Comper Swift advert from 1932.

Figure 16, opposite: A Lea Francis car typical of those made at Hooton Park in the 1950s.

The Return to Military use for the Second World War

In September 1939 R.A.F. 610 Squadron was mobilised and then sent to R.A.F. Wittering for final training. At the same time, Martin Hearn obtained a contract from the Ministry of Aircraft Production to repair many Avro Ansons, and later De Havilland Mosquito fighter-bombers. As No. 7 Aircraft Assembly Unit, the work also included the assembly of various types of British aircraft and also American aircraft which arrived via Liverpool docks. The first helicopters used by the Allies were assembled and tested at Hooton towards the end of the war. A U.S.A.F. pilot who landed here occasionally was Thomas P. Williams, who will figure later in the story of Hooton Park.

The racecourse grandstand burnt down accidentally on 8th July 1940. The fire caused the complete destruction of a 1919 Issotta Fraschina motor car with a magnificent "coupe de ville" body.[12] The true cause of the fire was never really determined.

In 1941 the grass airfield was transformed to include a 6,000 foot concrete runway – one of the longest in Europe at that time. As aircraft became obsolete they were sent from all over the country to be scrapped at No. 100 Sub-storage site at Hooton. The end of the war brought a decline in work for Martin Hearn and the company then had to seek alternative work: buses were repaired, armoured cars were overhauled and Slingsby gliders manufactured.

Various Post Second World War Uses

In 1947 Martin Hearn's company was re-named Aero-Engineering and Marine (Merseyside) Ltd. However, Hearn himself was no longer involved. He went into partnership with Lily Belcher and ran the Eastham "Glider Club" as a successful private hotel for some 25 years. This was adjacent to the airfield at its north western corner facing Eastham Parish Church. The company took on a number of varied contracts, including manufacturing Kirkby Gliders and had completed three of these and made fuselages for a further fifteen before the contract was cancelled by the incoming post-war Labour government. Seeking alternative work using their skills and know-how in building ash frames which would survive both wet weather and changing temperatures they built "woodie" shooting brake bodies for both Rolls-Royce and Lea Francis cars. They also rebuilt over 350 bus bodies and later they serviced Canadair Sabre jet fighters for the Royal Canadian Air Force. The engineering company survived until 1955.

Wing Commander "Wilbur" Wright opened a flying school at Hooton and later a gliding club was operated from the northern end of the airfield. 610 Squadron Royal Auxiliary Air Force returned to Hooton Park in 1946 after valiant war service latterly flying Spitfires in mainland Europe. 663 (AOP) Squadron was reformed at Hooton Park in 1949 using Auster spotter aircraft. 610 Squadron received Meteor twin jet fighters in 1951 and 611 Squadron (West Lancashire) relocated to Hooton from Woodvale to use the longer Hooton runway required for this type of aircraft. The three squadrons operated as Royal Auxiliary Air Force units from the airfield until all Auxiliary flying squadrons were disbanded in March 1957. At this point the R.A.F. Station was closed and all flying at Hooton Park ceased.[13]

After the Closure of the Airfield

The closure of the aerodrome was not the end of the story for Hooton Park. In 1959 it became the site (intermittently) of the north's biggest agricultural show (The Cheshire Show) until 1977 and the runways continued to be used by Shell Research for testing cars at high speed. In 1959 Hooton Park came to the attention of Vauxhall Motors Ltd. as a possible site for expansion of their production activities.

[12] Typescript notes by Brian Ramsay, CRO Ref: D 6686

[13] *op. cit.*, Peter Richardson, *Hooton Park a Thousand Years of History*, Hooton Airword, South Wirral, 1993 and Wikipedia (http://en.wikipedia.org)

What is Left of Hooton Hall Now?

In 1849 when Hooton Hall was incorporated into the enlarged hall, pillars from the 1778 hall were removed to allow the new grand portico to be constructed. These were built into a shippon at Park Farm in Hooton and can still be seen from Hooton Lane on the left-hand side just before Rivacre Road is reached. They are the only known remains of the 1778 hall.

Eight pillars from the new (1849) portico were bought by the renowned architect Clough Williams-Ellis sometime in the 1930s. It was not until nearly thirty years later that he incorporated them into the front of the "Gloriette" at his Italianate village, Portmeirion, on the edge of the Snowdonia National Park in North Wales. He had mislaid them in the intervening period and was embarrassed to find them buried in a garden in his village.

Standing near the site of Hooton Hall today presents a very different scene here from August 1914 when the air ministry requisitioned the site for the Royal Flying Corps. Of the three hangars built during the First World War the roofs of two are collapsing. With regard to the numerous wartime huts some are in use by the Griffin Trust who run a small transport museum, while the others just have their concrete floor slabs with some scattered Marley tiles and have small trees and scrub growing around them. Some have the remains of walls and one serves as the Vauxhall gardening club shop.

Turning to the once proud parkland, the racecourse and grandstand were covered at first by a concrete runway and perimeter track, but now by a vehicle storage compound and despatch offices. The south eastern end of the park with its ponds and small woods is now the giant Vauxhall car plant. Where the hall actually stood, the land has been excavated to make it level with the car assembly building.

There were four statues of goddesses at the top of the portico of the 1849 hall and a head from one of these statues now lies in a private garden near Ellesmere Port. A Stanley family crest from Hooton Hall can be seen on the front wall of the Stanley Arms pub in nearby Eastham village.[14] Oak panelling from the hall lines the Star Chamber in the Leasowe Castle Hotel at Moreton, Wirral.

Cheshire Record Office has a collection of documents relating to the Stanley families and the Naylor estates. Amongst these is a partial inventory of the hall from 1849. This record includes details of many of the main rooms, servants' quarters, outbuildings (including a brew house and a mash house) together with a list of all plants in the gardens.[15] Another bundle of legal documents relate to legal actions between Mr. Naylor and the Board of Trade concerning the right to the salvage of wrecks on the foreshore. They also include details of his dispute with the Manchester Ship Canal Company concerning ownership of the land to be used for the proposed ship canal.[16]

On the ground very little evidence of the once grand country estate can be seen, although the entrance gates still stand next to St. Paul's Church at Childer Thornton on the A41 Chester Road. In the early years of the Second World War the final traces of the racecourse were obliterated when the land was drained and runways were laid out by the Air Ministry. Sometime in the 1930s most of the rubble from the house was removed to raise the road over the Shropshire Union Canal at Backford when the A41 was improved.

Occupation of Stud Farm is recorded for the last time in the 1967 electoral register. It no longer had any farmland and was demolished during the landscaping of the area surrounding the Vauxhall plant when the M53 was constructed.[17]

Members of the Hooton Park Trust, recently formed to care for the three Belfast Hangars, have an interest in the history of the hall and have published a far more detailed work on the airfield's history, *Hooton Park: a Thousand Years of History*.

[14] Norman Ellison, *The Wirral Peninsular*, Robert Hales & Co., 1955

[15] CRO Ref: D 2305/6

[16] CRO Ref: D 2305/11

[17] *Electorial Register Wirral*, 1967, CRO Ref: CCR 1/289

Hooton Park in 1959

The main airfield runway can be seen running diagonally across the picture and the perimeter track virtually surrounds the airfield. At the top can be seen the river Mersey and Bowater's paper mill is near the top right-hand corner. North Road which comes from Ellesmere Port can be seen passing Bowater's and it runs towards the oil storage tanks at centre left. North Road doubled as part of the perimeter track for this length. The perimeter track turns left towards the lower edge of the picture and then turns left again towards the three Belfast hangars before skirting around the patio of the old Hooton Hall. It then runs parallel with Rivacre Road and passes Stud Farm before joining the end of the main runway.

Figure 17, opposite: The Goddess's head.

Figure 18: Hooton airfield Summer 1959.

Ellesmere Port: a Very Short History

Ellesmere Port is a relatively new town, so-named as it grew up around the basin for a canal, linking Ellesmere in Shropshire with the river Mersey, which opened in 1795 as far as Chester and was soon to connect Ellesmere, Shropshire to the river Mersey at Whitby Locks. These are located in the old village of Netherpool. This canal was part of the "Grand Crossing" scheme to inter-connect the four major English rivers, the Mersey, Severn, Thames and the Trent. The new town of Ellesmere Port grew around the basin with its flour mills and warehouses. The railway station opened in 1863 and the Urban District of Ellesmere Port and Whitby was created on 1st April 1902. The Civil Parishes of Overpool, Netherpool, Stanlow and Great Stanney were added to the Urban District 1st April 1910.

20th century industrial development centred on the oil refining, chemical and paper industries. Extracts from the 1961 and 1971 census for England and Wales reveal that while the population grew from 44,717 to 61,637, the number of people working in Ellesmere Port but living elsewhere grew from 9,610 to 26,130 persons. These figures reflect the huge increase in employment in the borough.[18] Vauxhall employment at Ellesmere Port during this period increased from nil to nearly 12,000.

[18] Peter J. Aspinal and Daphne M. Hudson, *Ellesmere Port: The Making of an Industrial Borough*, Ellesmere Port Borough Council, 1982

Vauxhall Motors Ltd. at Luton and Dunstable

Where did the Griffin come from? The heraldic Griffin emblem has long been associated with both Luton and Vauxhall cars. It originated in the 13th century when it was chosen as his coat of arms by Fulk le Breant, a soldier of Norman descent, following his appointment as a Sheriff and being given the Manor of Luton by King John.

Figure 19: A Griffin.

Figure 20: A Wyvern.

Is it a Griffin or a Wyvern? Exactly what is the fabulous beast on Fulk le Breant's coat of arms and today's Vauxhall logo? Vauxhall seemed unsure of themselves when they called a series of 1950s model Wyverns and in the 1990s called their Luton HQ Griffin House. The dictionary defines the two contenders as follows: a Griffin (or griffon) is a fabulous animal typically half eagle and half lion, whereas a Wyvern is a fabulous animal usually represented as a two legged winged creature resembling a dragon.

Look at any Vauxhall badge over the last century and it's certainly an eagle on top (just see that beak and eyes and compare to the Welsh dragon) but it's represented as a two legged creature as our friend is standing on a heraldic device called a bridge. The clincher, however, is the bushy end to the tail and the fur feathers on the front legs which shout lion! So he is a Griffin and Vauxhall should be ashamed for introducing the interloper Wyvern which has no place in Vauxhall history aside from being the name of some bland four cylinder cars built in Luton during the 1950s.

When Fulk le Breant married a wealthy widow he acquired his wife's property on the bank of the Thames at Lambeth. The house soon became known as Fulk's Hall. Over the years the name was corrupted to Foxhall then Vaux Hall and eventually to Vauxhall where the grounds became the famous (some say infamous) Vauxhall pleasure gardens in the 18th and early 19th century. This area of London is still known as Vauxhall.

There is a story that a visiting Russian delegation came to Britain to inspect the construction of the new-fangled railways in 1840. The British government, thinking they were on a spying mission, only allowed them to see Vauxhall Station in south London, which was then under construction. When told the name of the station, they assumed that it was the generic name for all important railway stations, and this supposedly explains why today the Russian word for a large railway station is Вокзал, which transcribes as Vokzal and is pronounced Vauxhall.

Vauxhall Motors Griffin logo has morphed over the 100 years of the company's existence and the one shown on this book's title page dates from about 1963.

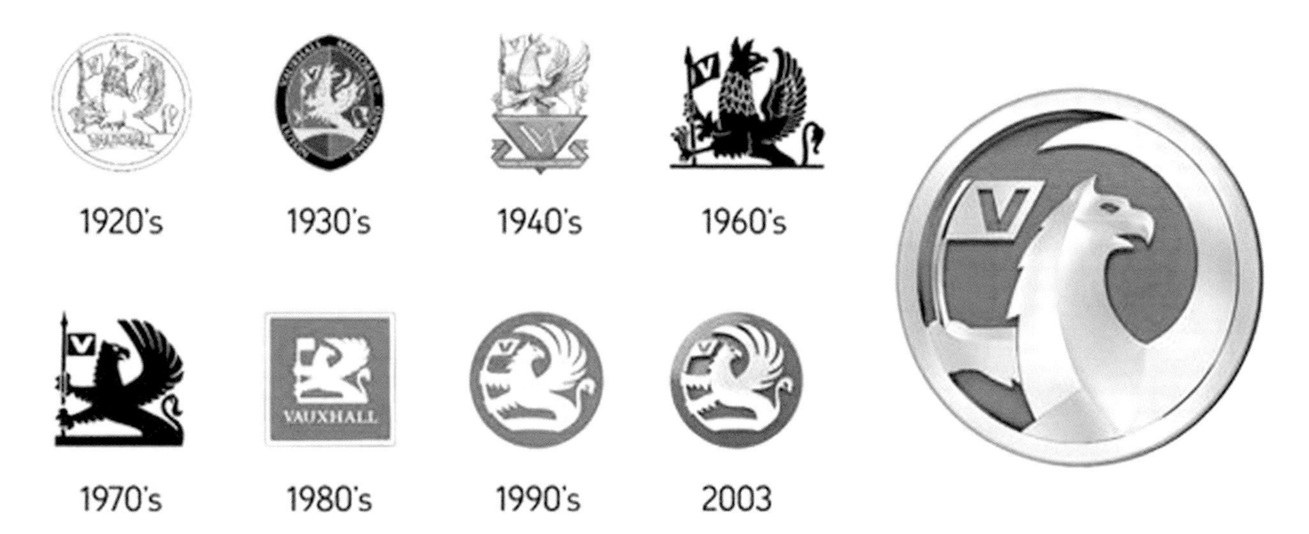

Figure 21: The Griffin changes over 100 years.

Figure 23: F. W. Hodges, chief engineer *circa* 1909.

Vauxhall's roots were not, as was often the case with early car companies, in the bicycle, sewing machine or carriage trades, but uniquely with steam engines for tugboats and Admiralty pinnaces (fast communications launches for use in harbour). The Vauxhall Iron Works had been founded by Alexander Wilson in 1857 at 90-92 Wandsworth Road in the Vauxhall area of South London. A Sainsbury's petrol station and supermarket car park now occupy the site. Today a wall plaque commemorates Vauxhall's origins. Just up Wandsworth Road from Wilson's site was the Queen's private railway station where Her Majesty would have boarded her London & South Western Railway train for journeys to Portsmouth and then on by ferry to Osborne House on the Isle of Wight.

Alexander Wilson & Co. are recorded in several London trade directories during the second half of the 19th century, as makers of cement testing machines, rock drilling and compressed air equipment makers at Vauxhall Iron Works, Wandsworth Road, Lambeth. Wilson must have built up quite a substantial company as the 1881 census shows him, aged 40, (but no Mrs. Wilson) living at 82 Jeffries Road, Lambeth employing 95 men and 24 boys.

Late in the 19th century, after a spell in receivership, Wilson left the company and it was renamed the Vauxhall Iron Works Company Ltd. Soon one of their engineers, Mr. F. W. Hodges, who had been an apprentice to Wilson, recognised the potential for cars and developed an interest in petrol engines.

Hodges persuaded the company to look at a few of his designs and soon a single cylinder petrol engine with two opposed pistons was tested in a river launch. Work continued throughout 1902 and by the summer of 1903 the company was confident enough to place on the market a single cylinder car which was priced 130 guineas (£138). In 1903 150 men manufactured 43 cars.[19] With this new venture the company soon outgrew their original "iron works" and were also renting several other workshops around the Vauxhall area.

[19] *op. cit.*, L. C. Darbyshire, *The Story of Vauxhall 1857-1946*, Vauxhall Motors Ltd., Luton, 1946

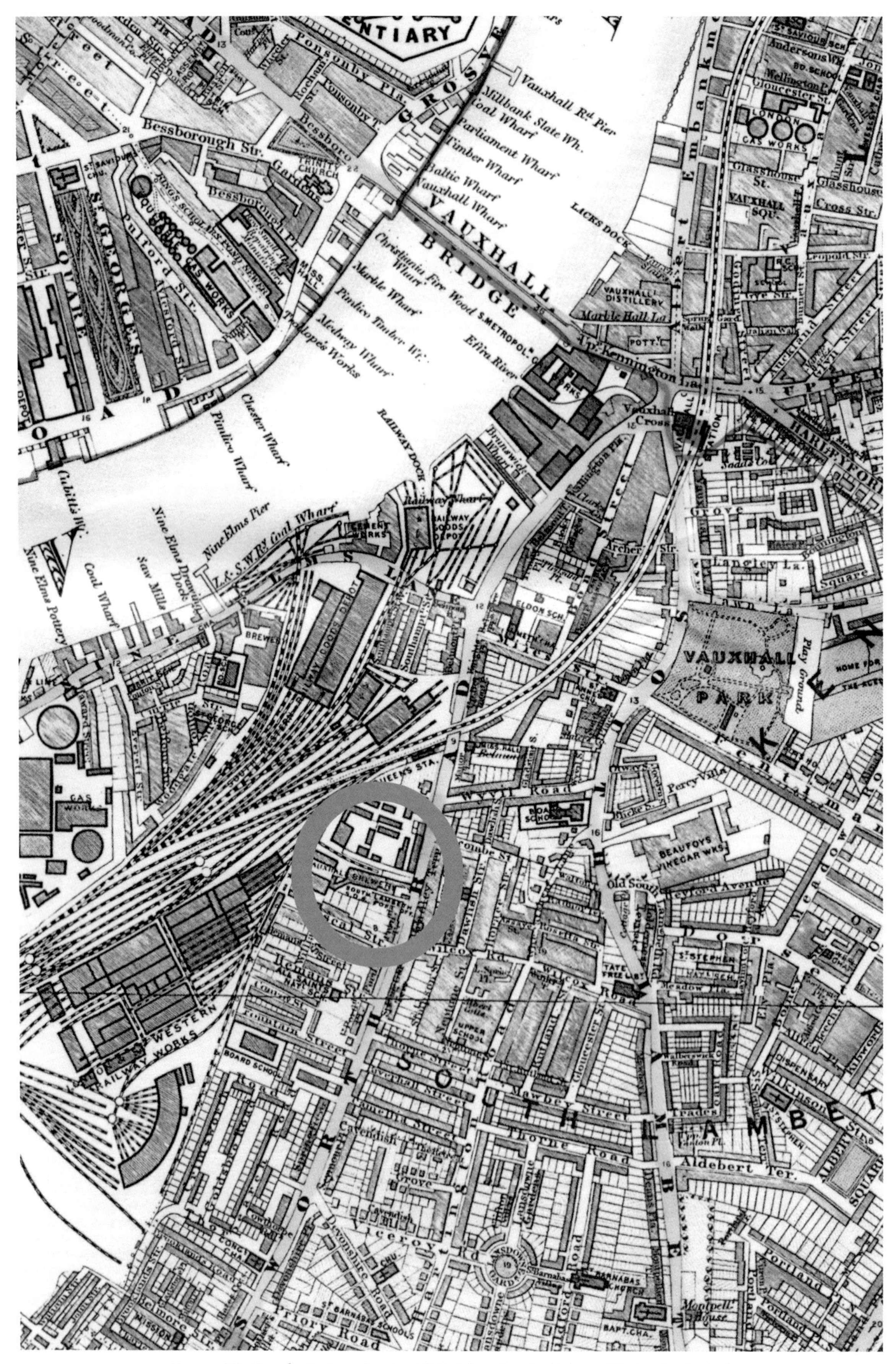

Figure 22: Vauxhall, London in 1893, The Vauxhall Iron Works is circled.

Vauxhall Comes to Luton

The town of Luton in Bedfordshire had long been the centre of a thriving hat industry, utilizing straw from the surrounding farm lands, but it was an industry in decline. In 1905 the Luton Chamber of Commerce was concerned at the high levels of male unemployment in the town due to this decline and they set up a "New Industries Committee" to encourage economic diversity. Vauxhall were enticed to Luton by the comparatively low wage levels, nominal trade union activity and low land prices.

Figure 24: A sketch of the factory *circa* 1906.

The site at Kimpton Road in Luton, then little more than a country lane on the outskirts of the town, gave them vastly more room to expand. The new factory, initially 6 ½ acres, faced Kimpton Road and had a siding from the Midland Railway line which ran from St. Pancras in London and then north towards Derby and beyond giving the company excellent road and rail links.

In 1906 Vauxhall Iron Works amalgamated with their Luton neighbours, West Hydraulic Co. and became known as the Vauxhall and West Hydraulic Engineering Co. Ltd. Hardly a snappy title by today's standards, nor apparently a happy relationship for in the next year, 1907, the companies separated. Vauxhall Motors Ltd. was formed to carry out the business of automobile engineers while the "Hydraulic" reverted to their original name and business. Although the firms remained neighbours until the 1960s they had no common interests.[20]

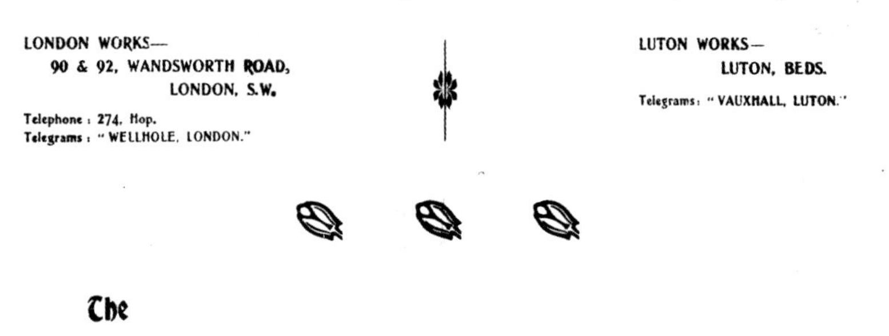

MOTOR & MARINE ENGINEERS.

6 H.P. CAR SPECIFICATION.

Figure 25: This pamphlet dates from about 1905, soon after the move to Luton. They still retained the London works and apparently had not yet installed the telephone at Luton. The car shown has tiller steering – a steering wheel was an optional extra!

[20] Len Holden, *Vauxhall Motors and the Luton Economy 1900-2002*, Boydell Press, 2003

Manufacturing methods in 1910

Virtually no records exist about production methods prior to the First World War but the following series of photos from 1910 show that production methods have become a little bit organised. Activities were departmentalised into specialist workshops although still operated on a craft production basis, and each man was a craftsman. Indeed any village blacksmith would immediately be at home in the forge department, shown below, using traditional tools and methods.

Figure 26: The forge shop.

Figure 27: Engine shop.

This picture of the engine assembly area seems to show each man building a whole engine, the four cylinder crankshaft looks as though it has five main bearings, a feature Ford were trumpeting as a new development in the 1960s.

Figure 28: The Erecting shop in 1910.
The picture shows the chassis being built up *in situ*. The chassis number 1231 on the nearest car seems to be a little bit inflated as by the end of 1910 Vauxhall had only built 736 cars since the first one in 1903.

This rather gloomy picture shows the area where the bodies were assembled onto the chassis. It is not known whether the bodies were actually assembled on the chassis or built up off line and put on as a complete unit. This is the first shot to show cars facing in a line so they could be pushed along on their wheels. Some chassis would have been sent to specialist coach works for their bodies to be made to the customer's requirements.

Figure 35: Car Assembly line *circa* 1920. Note the wheels without tyres running on the tubular track.

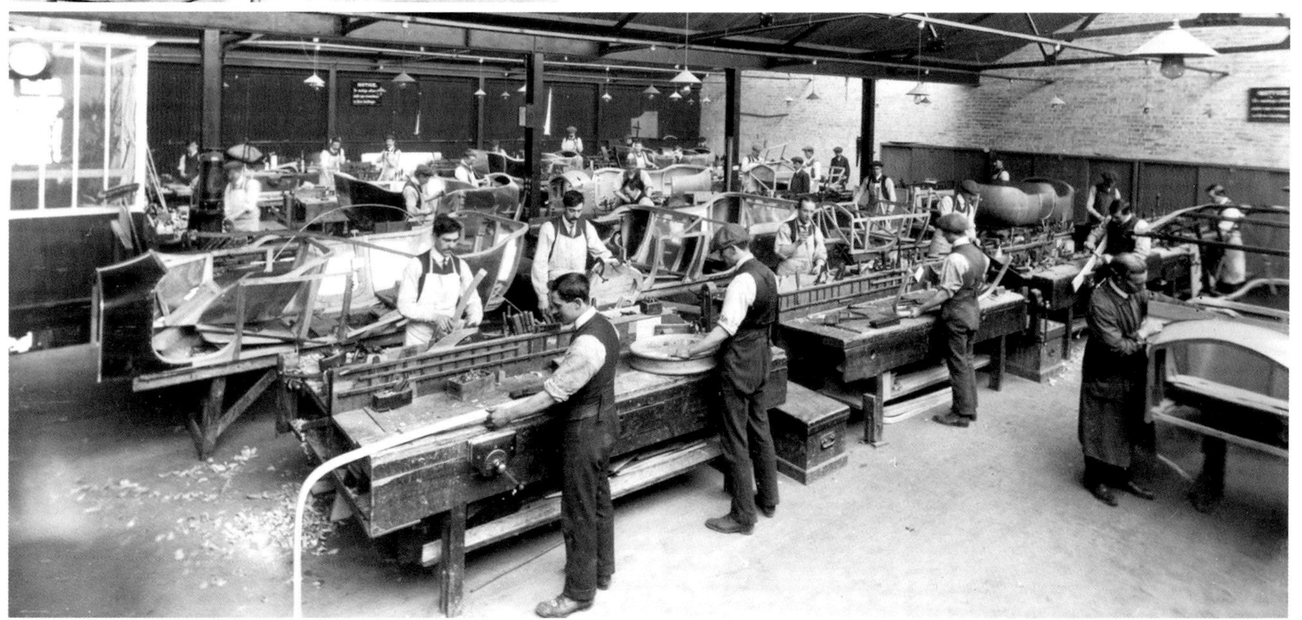

Figure 29: Body shop.
In this body shop picture each craftsman is at a traditional carpenter's bench and using hand craft methods of production.

Although the caption for figure 31 has been copied from the contemporary print in the Vauxhall archives, it looks more like a finished vehicle area, although why some cars appear to have no tyres can only be guessed at.

Figure 31: Erecting shop highway shows a Prince Henry coming out of G-Block.

Descriptions of most of the cars produced by the company have been well-recorded elsewhere but a couple of distinctly odd products have escaped the attention of many authors. About 1905/6 this impractical hansom cab was produced, but probably only in small numbers. One was kept by the company and used regularly to carry the cash from the bank in Luton to the factory on pay day, but none are known to have survived.

Figure 32: 1905/6 Hansom cab.

In 1923 the Engineering Department[21] produced six prototype motor cycles designed by Ricardo' Engineering Co. following that company's successful project for the Vauxhall Ulster TT car. These bikes had a four cylinder engine and shaft drive but the bike never went into production and the prototypes were sold to employees. Only two examples are known to have survived.

Until 1925 Vauxhall had a reputation for making high quality and often sporting cars. The epitome of these cars was the Prince Henry model produced from about 1910. Figure 34 shows a painting commissioned in 1965 for presentation to Percy Kidner who had been joint managing director of Vauxhall Motors from 1907 until 1928. Mr. Kidner is the driver in this fine picture by Michael Turner.

Figure 33: Vauxhall motor bike of 1923.

Figure 34: A Vauxhall Prince Henry taking part in the 1911 Swedish Trials. A painting by Michael Turner commissioned by Vauxhall Motors in 1965 and presented to Percy Kidner.

[21] Vauxhall Motors' vehicle design and development department has always been known as "Engineering"

General Motors Corporation

By 1925 the company was again in financial trouble. This time America's General Motors Corporation came to the rescue. GM executives had visited Britain with the intention of buying the Austin Motor Company, but criticism by the British motoring press[22] which resented a well-known firm falling into American hands and with dissenting (Austin) directors favouring a modest capital restructuring they avoided American acquisition.

Soon after the Austin negotiations broke down GM opened discussions with Vauxhall Motors. The Vauxhall board was more willing to entertain GM's offer of $2,575,291 for the company's ordinary share capital. The old ordinary shareholders were paid a £210,000 bonus and in British terms GM had invested a total of £510,000 to acquire the company.[23]

Over the next few years it changed its focus from hand-built, high quality sporting cars to the mass production of more popular cars with a diversity of models to capitalise on their market potential. In 1929 GM also acquired the German motor manufacturer Adam Opel A.G.

During the early 1930s Vauxhall started producing the Bedford range of light commercial vehicles although Chevrolet trucks from CKD kits had been assembled at Hendon since 1924.

Vauxhall soon expanded the Bedford range into a full catalogue of truck chassis on which independent body builders produced a wide range of lorries, vans and coaches. By 1950 Bedford had become the dominant UK truck producer.

Figure 36: This pair of Bedford coaches was operated by Gorman Bros. of Coatbridge, Lanarkshire in 1932.

[22] This was to be repeated in 1985 when HM Government encouraged GM to take over Land Rover and Leyland Trucks but the deal was scuppered by the press and local political pressures.

[23] Holden, *op. cit.*, p29

The Second World War saw Vauxhall Motors turn their production facilities over to the war effort and in addition to being one of the main suppliers of trucks to the armed services they designed (in 88 days) and then manufactured some 5,600 Churchill tanks. Utilizing their sheet metal stamping facilities Vauxhall produced literally millions of "tin hats", jerry cans and shell cases.[24]

One unusual design exercise during the Second World War was these inflatable trucks. They became part of a programme using decoy tanks and buildings etc. to trick the enemy into believing that our forces were stronger than they were in reality and to divert their attention from the genuine sites.

Figure 37: A decoy inflatable "truck".

Peace in 1945 saw the return to car production at Luton. By the mid-1950s the company had opened a new factory at Dunstable in Bedfordshire, just five miles away, solely to produce Bedford trucks. Such was Bedford's domination of the UK market at that time that they sold 50% of the trucks in the domestic market and 45% of the trucks exported from the U.K. were Bedfords.

In the 1950s Vauxhall, Opel, AC Delco, Frigidaire and other companies owned by GM became part of General Motors Overseas Operations (GMOO), although each remained separate companies and for legal reasons Vauxhall Motors Ltd. remained a UK company registered at Companies House until 2008 when it became GMUK Ltd.

During the late 1950s the need for further export trade and an ever-expanding home car market led the company to seek another new factory site near to its Luton plant. Over the years the plant had become hemmed in by housing estates, the new Luton Airport and the railway line. Although some land was available on the Dunstable site it was not sufficient for the company's ambitious expansion plans.

[24] *op. cit.*, L. C. Darbyshire, *The Story of Vauxhall 1857-1946*, Vauxhall Motors Ltd. Luton, 1946

Employment Conditions at Vauxhall, Luton

During the 1930s Vauxhall's employment levels were increasing rapidly, driven by the expansion of production volumes arising from the change to the manufacture of popular, lower-priced cars. Many of these new workers were drawn from the valleys of South Wales, Scotland, Ireland and the north of England, all areas hit by the depression prevailing at the time. They were attracted by the prospect of steady employment in Luton. These newcomers soon settled in Luton and after the war Vauxhall Motors adopted a policy of recruiting the family members of established employees. There were many families with siblings and two or even three generations working at Vauxhall. Many workers reached 25 years service, when they were given a gold watch. Some even stayed for 40 years or more when a fine clock was often presented. There is one family (the Walters) recorded where the grandfather was originally employed at Wandsworth Road and he moved to Luton in 1905 where his son worked for many years. Later his grandson started his working life at Luton, but transferred to Ellesmere Port in 1965 from where he retired in 2000. All three generations and their various siblings spent much of their lives working for Vauxhall. The Walters family also ran the newspaper kiosks at the factory gates, doing a brisk trade every morning before some of them came into work on the assembly lines.

Most workers travelled to work by bike, bus or on foot. The Vauxhall factory at Luton expanded over the years and by 1950 occupied land on both sides of Kimpton Road. This was one of the main roads out of Luton and was also the main route to the airport. In the late 1950s each side of Kimpton Road was lined with bus stops. At the end of each shift rivers of workers pouring from the factory gates could be seen rushing to catch their buses or on their bikes weaving their way through the throng of pedestrians, buses and cars.

As an employer Vauxhall Motors, led by Sir Charles Bartlett between 1929 and 1953, took an enlightened attitude to labour relations. During the Second World War they had set up the "Management Advisory Committee" (MAC). This was a joint consultation/discussion body between management and workforce representatives to resolve issues before they got out of proportion. The elected workers' representatives were released from their normal duties to act as friends to their constituents. They would listen to their work or welfare problems, often helping them to find solutions, or where there was no ready solution, explaining the company's policies and views.

Luton was outside the normal circuit of trade union militancy, clear of London and too far from Birmingham to earn the attention of practised agitators. Until the early 1960s it remained relatively isolated from the trade union influences of these two areas.[25] In general trade union activity before the 1960s was fairly low key. Although Vauxhall recognised the trade unions who, although they had negotiating rights, worked alongside the Management Advisory Committee. This committee met formally every month but the "MAC man" had open door access to senior management to resolve problems before they got out of hand. Although trade union membership was allowed it was not encouraged and membership was low. Therefore the unions played the minor role to the Management Advisory Committee in discussions with management even though some MAC representatives wore two hats if they were also active trade unionists.

Vauxhall had always had a policy of promotion from within the company. Indeed Sir Reginald Pearson started his career as a lathe operator in 1919 and was promoted successively from a machine shop foreman to deputy chairman of the company, the position he held when he retired in 1963. There was virtually no recruitment of staff and hence influx of ideas from other industries nor even other motor manufacturers. The perceived wisdom was that Vauxhall was big enough to grow its own management team, the only exception being a few senior managers and directors who were transferred from other divisions of General Motors. GM even had its own university, the GM Institute based at Flint in Michigan, USA. Each overseas division of GM sent promising ex-apprentices to Flint for a two year management training course. On their return to their home plant they were often fast-tracked to senior management positions.

Management gurus would probably describe this policy as incestuous, and in one or two areas you had to be a catholic or a freemason to be promoted. It was not until the 1970s that university graduates were recruited directly to junior management roles such as foremen or staff positions.

[25] Graham Turner, *The Car Makers*, Eyre & Spotiswoode, 1963

A study in 1968 described Vauxhall as "The industrial soul of Luton". It was a magnet for workers from other parts of Britain and the Indian sub-continent. Employees were paid £5 for each new recruit that they introduced and wages were 50% above the national average.[26] One former employee recalls a range of temporary buildings on the road to Luton Airport to provide living accommodation for new employees during the years of expansion 1955-60.[27] Employee benefits included a fully-equipped medical surgery with a full-time doctor and nurses, a rehabilitation centre for workers to recover from sickness or injury in a protected environment, two or three weeks paid holiday depending upon length of service, and a travel department specialising in holiday bookings. These holidays usually had to be taken during "The Vauxhall Shutdown" and it affected the town so much that many shops including Marks & Spencer also closed for the holiday weeks.

There were extensive canteens and a large sports ground with a clubhouse, all run by the Recreation Club to which employees automatically belonged. The Recreation Club had many sections covering activities from amateur dramatics, male and female voice choirs, an orchestra, photography, small bore rifle shooting and the full range of sporting groups who used the extensive Brache Estate sports ground. There was even a firewood rationing scheme to enable employees to obtain unwanted wood off-cuts free of charge. The Welfare Department was able to hire out various pieces of equipment to employees with short-term needs. For example a bedpan could be rented for sixpence per week. Wheelchairs and crutches could also be hired.

Figure 38: Employees in the "Rehab" workshop regain their strength and dexterity on light assembly work without the pressure and stress of a production line.

[26] Goldthorpe J., *et al.*

[27] J. M. Bates interview, February 2010

Why Did Vauxhall Expand in the North?

The National Economic Scene in the 1950s

While the economy generally was booming during the 1950s (Harold MacMillan's 1957 memorable speech at Bedford included the phrase "You've never had it so good") there were pockets of unemployment away from the prosperous south east and the industrial Midlands. In 1959 the national figure for unemployment was 450,000 (1.5%). On Merseyside unemployment was 25,000 or 3.5% of the local workforce. Because of this regional problem the Government began encouraging the motor industry to move to designated "Industrial Development Areas". These were defined as having unemployment levels in excess of 3.5%. They included Merseyside, the Tyne & Tees areas, Cumbria, Cornwall, the Forest of Dean, South Wales and Central Scotland. Merseyside was by then suffering from the decline of the Lancashire cotton industry, the docks and the ship building industry.

In 1957 the AC Delco Division of General Motors, who had their main factory in Dunstable, were assured that there would soon be customers nearby and were given Government subsidies. They set up an automotive electrical components factory, at Kirkby on the outskirts of Liverpool. About 1,500 local people were employed at this plant.

By 1959 the new industrial development on Merseyside was gathering pace. On 3rd May British Motor Corporation started work on the new Fisher Ludlow domestic appliance factory at Kirkby. Later that year it became known that the Board of Trade was having discussions with the "Big Five" motor manufacturers who all wanted to expand their production facilities. By the end of 1960 Rootes Group had announced that it would build a new car plant at Linwood near Paisley to the south west of Glasgow, British Motor Corporation would build a tractor plant at Bathgate east of Edinburgh and Ford, Vauxhall and Standard Triumph had all announced plans to open new manufacturing plants on Merseyside. Vauxhall Motors had looked at sites recommended by the Board of Trade in other designated industrial development areas such as Bristol, Newcastle upon Tyne and Clydeside. However, none of these fully matched the company's needs.[28]

A major inducement offered by the Board of Trade for investments in Industrial Development Areas was a 40% grant towards the cost of new plant and machinery and about 25% towards the cost of buildings. In a manner of speaking, the coming of Vauxhall to Merseyside was a strange repetition of the economic situation at Luton in 1905 and the action taken then to overcome the problems.

Other factors which impacted upon Vauxhall's plan to open a factory on Merseyside are reflected in extracts from newspapers and magazines of the period. Prior to 1959 the Cheshire Agricultural Society had held its annual show on the Roodee in Chester but it had outgrown that venue and in 1959 it moved the show to Hooton Airfield. This had easier road access and the area available for the show and car parking was far larger than the Roodee could offer in the city centre.

10th June 1959, *Chester Chronicle* (County Edition)

Record first day attendance at the Cheshire Show which was staged for the first time at Hooton Airfield, 36,110 people attended on Wednesday, the first day, to make it certain the last year's overall attendance of 42,514 at the Roodee in Chester would be topped. 8,000 cars attended on the first day whereas only 1,500 could park on the Roodee.

23rd May 1959, *Chester Chronicle*

200 Liverpool overspill families to move to Ellesmere Port next month, these are the vanguard of 20,000 people expected to arrive over the next 10 years. The cost of the housing is shared by Ellesmere Port Council and Liverpool City Council.

This policy may seem to have had the potential for creating a huge pool of unemployment in the borough, but one condition imposed by the borough's scheme was that anyone moving to Ellesmere Port must either have a job or the offer of one in the borough.[29] Another factor which would have an effect on unemployment levels in Ellesmere Port was that the children born in the post-war baby boom of 1946 to 1949 were now beginning to enter the jobs market.

21st November 1959, *The Times*

Vauxhall's Consider Future Prospects.
Discussions are going on between Vauxhall Motors and the Board of Trade to ascertain the position if it was decided to embark on a further expansion programme.

A company spokesman emphasised that the talks were purely exploratory and no definite programme was under consideration.

In January 1960, the town clerk to Ellesmere Port Borough Council, Raymond Bernie, read in *The Financial Times* that the major motor manufacturers were looking for expansion sites away from their traditional bases. He immediately wrote to all the "Big Five" manufacturers inviting them to visit Ellesmere Port to view potential sites. Only Vauxhall responded positively and on their way to a meeting with the council three Vauxhall executives, Sir Reginald Pearson (Deputy Chairman), Tom Williams (Project Manager) and Eric Fountain (Chief Factory Layout Engineer), came to Ellesmere Port from the A41 at Eastham via Rivacre Road where they spotted Hooton Park Airfield. After one look at the desolate airfield, the Vauxhall men decided it was the ideal location, having potentially good road access and being fairly level. There also appeared to be rail links close by and the opportunity of a sea route for transporting finished vehicles to the south of England.[30]

Although the company were shown several other sites in Cheshire and other parts of the UK they seem to have quickly focused their attention on Hooton Park for several reasons. It was adjacent to a future motorway, giving good access for employees, suppliers and finished vehicle deliveries, there was the nearby access to ocean-going shipping to and from North America, there were no nearby housing estates to suffer from noise pollution, the site was fairly level and there was plenty of scope for future expansion.

Other locations had been considered notably Glasgow, the Tyne Tees area, Cardiff and Gresford near Wrexham. However, all this was not to become public knowledge for a few months yet.

The preliminary planning process

On 25th February 1960 Vauxhall Motors Ltd. was granted an Industrial Development Certificate by the Board of Trade to develop land within the boroughs of Ellesmere Port and Bebington for the purposes of the manufacture of motor vehicles.[31]

[28] *The Times* newspaper, various dates

[29] Interview with Mr R. J. Bernie, January 2009

[30] Ray Bernie

[31] CRO Ref: LBE 7440/1/2

There had been much speculation and circulation of rumours for the next few weeks until the company made an announcement to the press regarding their plans for the long-term future. Extracts from the Vauxhall house magazine are shown below:

> **March 1960, *Vauxhall Mirror***
>
> ***Vauxhalls to Build Works on Merseyside.***
>
> ***Employment for 7,000***
> *A new commercial vehicle factory around Ellesmere Port…is planned by Vauxhall Motors Ltd. This was announced by Mr. P. W. Coplin Chairman and M.D. of Vauxhall in London last night. At the same time the company's expansion programme would provide additional buildings in the Luton and Dunstable area which would be devoted entirely to car plant. The first stage will cost about £30m and have an area of 2,500,000 sq.ft. and eventually employ about 7000 people.*
>
> *He said that they would have preferred to expand in the Luton and Dunstable areas but after discussion with the Board of Trade they would adapt their plans to the Government's policy of steering industrial development to areas of under-employment. More than thirty sites have been visited in areas of under-employment. The area around Ellesmere Port was considered to be the most suitable. The firm proposed to construct a factory for the manufacture of Bedford commercial vehicles – both trucks which are now made at Dunstable and light vans at present produced at Luton. Construction would be in two stages. Stage one which covered the erection of buildings to house a press shop etc. and would take two years to complete when it would employ an estimated 3,500 people. Stage two which covered further development of the site into a highly integrated commercial vehicle plant. This stage would entail the construction of further buildings to which plant and machinery would be transferred from Luton and Dunstable.*

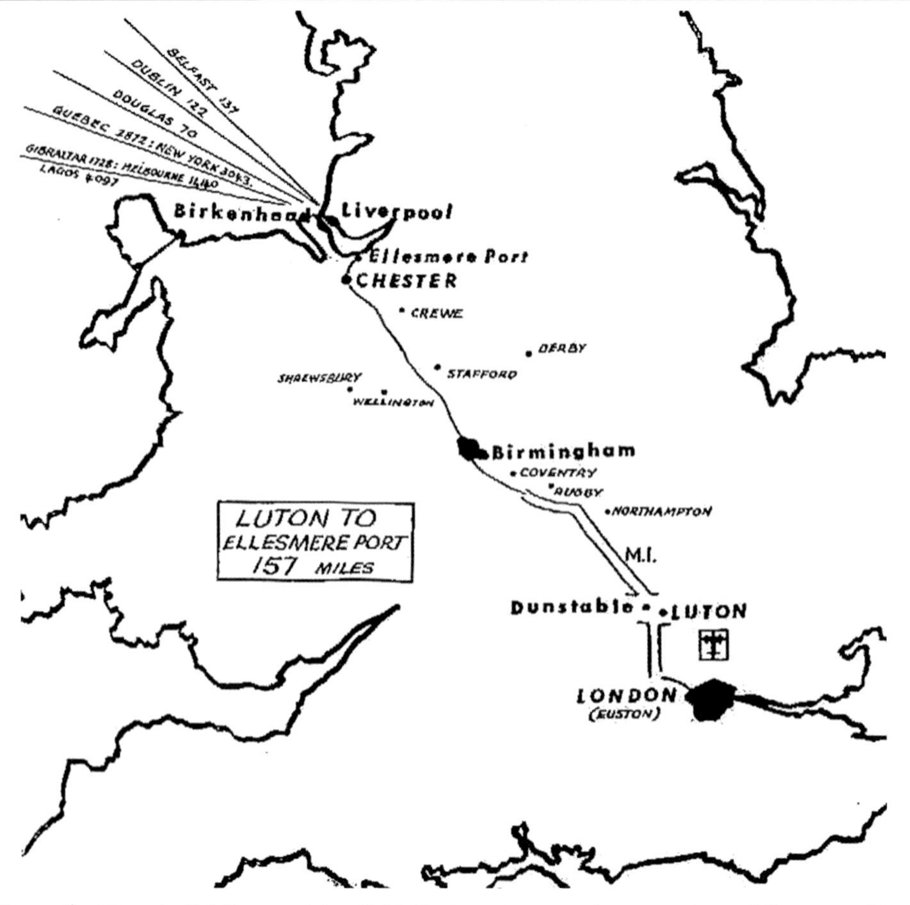

Figure 39: A map from the Vauxhall Mirror of April 1960 showing employees where Ellesmere Port was in relation to Chester and Liverpool.

At this stage five sites in Wirral had been investigated by the company and the actual preferred site for the new factory may have been decided upon but it certainly was not disclosed. Further articles went on to tell employees about Liverpool and Chester.

> ***Liverpool, the fourth city and first seaport of Great Britain,*** *is in Lancashire on the north bank of the Mersey, 3 miles from the sea. The city owes its prosperity to the impulse of the cotton trade at the end of the 18th century, through the port pass enormous imports of wheat and cotton and export of cotton goods. The population is estimated at 773,700 in 1956 and they are well provided with first class entertainment for Liverpool not only does Liverpool possess three theatres and 11 cinemas, she also has one of the finest concert halls in Europe, the Philharmonic Hall. Sportsmen (and women) are well catered for with football clubs and golf courses, skating rinks and 75 bowling greens. There are Association football matches at Anfield and Everton and Goodison Park*[32] (sic) *(the finest football stadium in the Provinces, housing 80,000 under cover), with horse racing and motor racing at Aintree. In fact with speedway racing and greyhound racing, 24 swimming baths, Russian baths, sun ray baths and many other remedial baths, the only complaint the population must have is that there isn't enough time for everything!*

> **3rd March 1960, *The Times***
>
> ***Car Firm's "Struggle"***
> *Sir Reginald Pearson, deputy Chairman of Vauxhall Motors said today that it was only after a struggle with the Board of Trade that the firm agreed to set up a factory at Ellesmere Port. It will put up our costs to move away from Luton and Dunstable where we have been part of the local life for over half a century. We shall have to start from scratch in an area where we are not an established firm – but merely a large industrial company from somewhere in the South of England...Proposals we submitted* [for expansion much nearer to Luton] *were turned down because – in the Government's view; they did not contribute enough to the employment position in "designated areas"*
>
> *Sir Reginald said that in the end the firm had agreed to the Government proposals only on condition that certain expansion was allowed in Luton and Dunstable.*

As the *Liverpool Daily Post* pointed out in an editorial, the Board of Trade had no powers to force companies to expand into the designated Industrial Development Areas, they could just refuse permission for expansion in other areas. The only alternative for industry was to relocate abroad!

> **5 March 1960, *Chester Chronicle***
>
> ***New Vauxhall Factory No Threat to Cheshire Show?***
> *With the announcement last week that Vauxhall Motors Ltd were to build their £30,000,000 factory in Ellesmere Port there was speculation as to whether Hooton Park might not be the chosen site, and fears were expressed about the future of the Cheshire Show on the old airfield. It is now believed that Vauxhall will site it to the south east of the town on the banks of the Manchester Ship Canal* [at Ince Marshes].
>
> *Hooton Park formerly belonged to the Trustees of the R. C. Naylor Estate and was acquired by the Air Ministry in the early days of the war. It was confirmed that discussions were going on for the return of the airfield to the trustees. It is understood that the Air Ministry are obliged to return the land to the original owners.*

[32] *n.b.* During the 1959-60 season Everton were in 15th place in Division 1 and Luton F.C. were in last place; Liverpool F.C. were in 3rd place in Division 2. During the season Bill Shankly was appointed manager of Liverpool F.C. and by 1963 Liverpool were top of Division 1 while Luton F.C. were near the bottom of Division 3 – how fortunes change!

28 April 1960, *Vauxhall Motors Annual Report for 1959*

Chairman's Comments
Total production for the year was 246,085 vehicles, 41% higher than 1958 very close to the quarter million vehicles for which the plant in its present capacity was laid down... We have been working on plans for further expansion to increase the capacity by 100,000 vehicles per year. In accordance with the wishes and policy of H.M. Government, we have agreed that part of this expansion shall be located in a designated area of under-employment and negotiations are in progress for the acquisition of a site on the south of the Mersey

The first evidence of any approach to the County Council, found in the Cheshire Record Office, is an Outline Planning Application from Vauxhall Motors dated 28 April, 1960.[33] This was for planning permission to carry out industrial development in the Ellesmere Port area for the manufacture of motor vehicles. Operations would include production, storage, research, welfare, administration and other incidental uses. The factory would cover a total area of 2.85m sq.ft.

30 April 1960, *Chester Chronicle*

Vauxhall Factory Site Speculation
There is still much speculation of a site near Ellesmere Port for Vauxhall Motors Ltd.

They had been offered a site at Stanney, [to the south of] *Ellesmere Port but the company were believed to prefer Hooton Airfield which is part of the Wirral Green Belt. The company were reported to be having discussions with the County Council, and the Municipal Councils of both Ellesmere Port and Bebington as the airfield lies astride the borough boundary. As the airfield is defined as green belt this would have to be relaxed and it is believed that Bebington Council are willing to do so.*

A council spokesman said that the company were likely to seek outline planning permission but until the details had been received no further statement could be made.

On 4th May 1960 the company again wrote to Cheshire County Council outlining the estimated stages of development of the plant. Stage 1 would be completed in 1963, stage 2 between 1963 and 65 and stage 3 (and probable further stages) at unspecified times. Stage 3 was only an estimate and would be dependent on product and market needs. The company pointed out that the total development still left acres of land undeveloped, which was in keeping with the Company's intention of acquiring an "open-ended" site with adequate room for long-term expansion.[34]

An undated drawing (Fig 40) appears to be associated with this letter and shows a building layout very different to the one which was ultimately implemented and seems to show what would have become Europe's largest commercial vehicle factory.[35] It shows a total planned floor area of 7.383 million square feet (equivalent to nearly 1,500 football pitches). Although the use of each building is not identified the drawing shows how the plant would be built in three phases. The layout is completely different from the factory later built, but it serves to illustrate Vauxhall's original long term plans for the Hooton Park site with possibly, a direct workforce of about 37,000 people.

On 16th May 1960 the Manchester Ship Canal Company (MSC) wrote to the county council outlining their objections to Vauxhall's application largely on the basis that some of the land was owned by MSC (although it had been used by the Air Ministry under requisition) and they (MSC) had long-term outline plans for its use.[36]

[33] CRO Ref: LBE 7440/1/1

[34] CRO Ref: LBE 7440/1/13

[35] CRO Ref: LBE 7440/2/38

[36] CRO Ref: LBE 7440/1/6

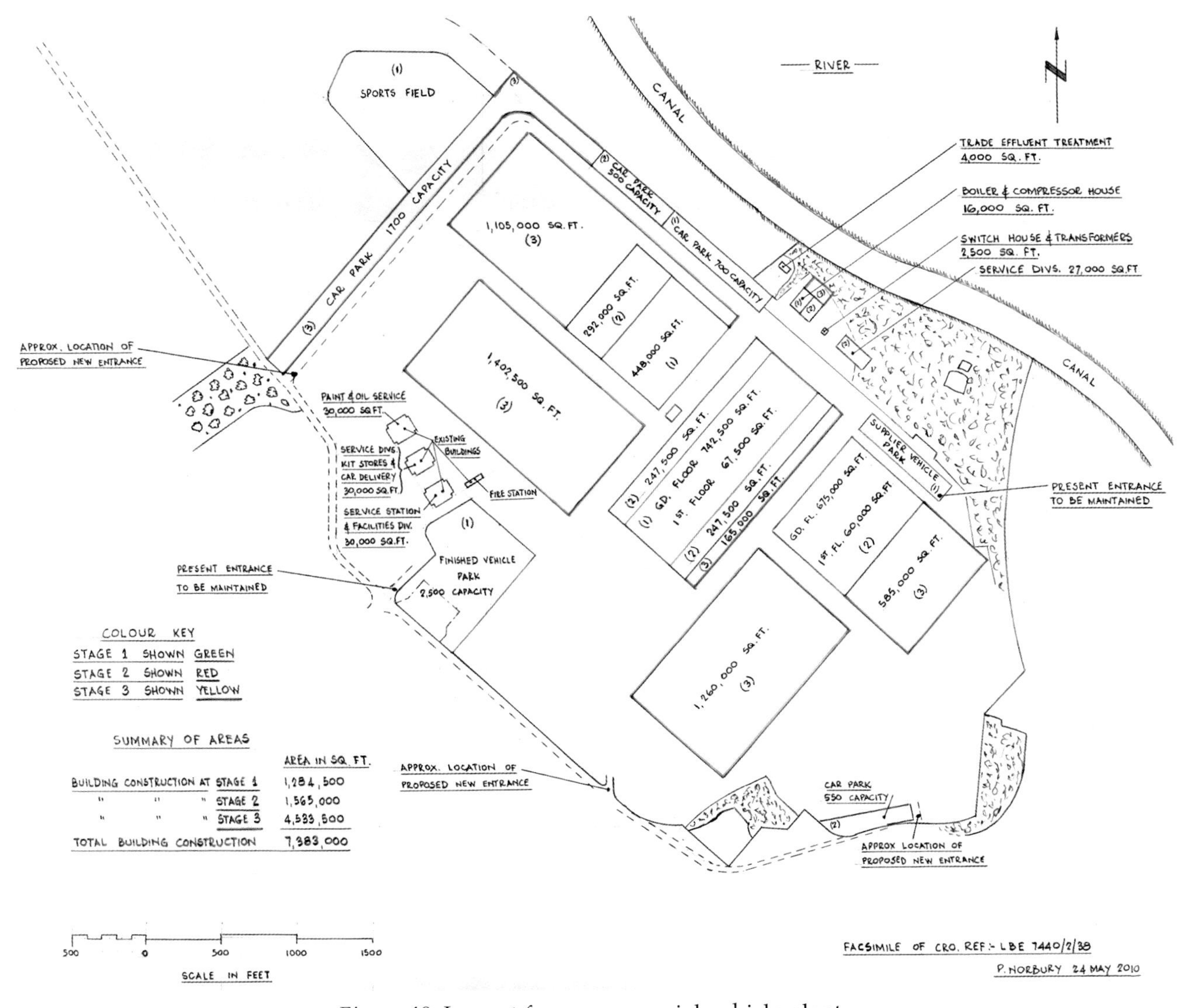

Figure 40: Layout for a commercial vehicle plant.

21st May 1960, *Chester Chronicle*

Vauxhall Want to Develop Cheshire Show Ground.
After considering a selection of sites in the Ellesmere Port area, Vauxhall Motors have applied for permission to develop about 400 acres at Hooton Park scheduled as a green belt area. The County Council is believed to be opposed to a factory erected in the green belt. Although consideration has not yet been given to the plans, the Ministry of Housing and Local Government has already ordered a public inquiry into the application to be held at County Hall on June 14th.

28 May 1960, *Chester Chronicle*

Cheshire Show: Still Hope of a Permanent Site.
Cheshire Agricultural Society is concerned that they will lose the newly found showground. A meeting of the Society's executive council held the previous week gave its officers powers to take any action and the matter was under discussion with the Society's solicitors. However, there was more than a remote possibility that a permanent show site was being provided on part of the former airfield.

Mr. P. Theaker of Bebington Council said that if Vauxhalls were allowed to develop about 113 acres would not be required. The society was encouraged by this news as the show required about 100 acres.

Ray Bernie, town clerk to Ellesmere Port Borough Council addressed the county council late in May 1960 and explained the problem that Hooton Park was provisionally designated as Green Belt. He said that large areas were under tarmac and concrete and that if it was to be declared Green Belt now, a future developer may appeal and have a good argument that it was not really "green". Under these circumstances the county council would probably lose the appeal and then would have less control over the development of the airfield than with the current proposal from Vauxhall.[37] As a result the county council appear to have dropped their "Green Belt" objections to the Vauxhall proposal.

Figure 41: The Cheshire Show, 9th June 1960, a few days before the public inquiry opened.

Cheshire County Council Give Tacit Approval to Ministry

On 3rd June 1960 Cheshire County Council wrote to the Ministry of Housing and Local Government saying that they had no objection in principle to the application but suggested a number of special provisions be included with any planning permission approval:

1. Any permission should be personal to Vauxhall Motors Ltd., and any area not developed by Vauxhall should not be available to other industrialists.

2. Permission should be for one comprehensive unit for the manufacture of motor vehicles.

3. Buildings etc. should be located to leave extensive areas of land open on the north west and south west of the site.

4. The open areas should be maintained as agricultural land or for other uses compatible with a Green Belt.

5. Land should be reserved for new and improved roads.

6. Tree planting and landscaping should be carried out to provide screening of the industrial development.

7. Steps should be taken to protect the amenities of nearby dwellings.[38]

[37] Interview with Ray Bernie, 13th January 2009

[38] CRO Ref: LBE 7440/1/5

The Public Inquiry

The Chronicle Saturday, June 18 1960

VAUXHALL MOTORS' PLANS FOR BIG WIRRAL FACTORY

£30 million - and work for 7000 by 1965

INQUIRY TOLD OF AERIAL SURVEY OF CHESHIRE SHOW

"Needs only about 64 acres"
says works engineer

At a local meeting conducted by Mr. M. B. Tetlow, an Inspector for the Ministry of Housing and Local Government at County Hall, Chester on Tuesday 14th June representations were made by Cheshire County Council, Manchester Ship Canal Company, Cheshire Agricultural Society, Vauxhall Motors Ltd., The R.C. Naylor Settlement, Kelvinator Ltd. and Lever Bros. Ltd. [Kelvinator and Lever Bros. produced refrigerators and soaps respectively]

Members of Vauxhall's management who attended included Mr. C. Stimpson, Assistant Co. Secretary and Mr. T. P. Williams, Works Engineer who would be in charge of the whole operation of building and transferring plant. [He became the first Plant Manager in 1962]

Outlining Vauxhall's application, Mr. Geoffrey Lawrence Q.C. said that the project was one of great importance to his clients, although the County Council and the two local authorities were not opposing the application in principal but had made one or two reservations or provisos.

Vauxhall said that they could not accept restrictions to their activities, for example if diesel engines were made for their trucks but to restrict them from selling these to other industrial users this would not be acceptable. Responding to objections from Wirral Green Belt Council the company could not accept that part of the site be left as a green belt until it was needed for building as a security fence would have to be put around the whole site.

Vauxhall could not wait until the future road pattern in the area had been determined by the County Council.

Regarding the Cheshire Show, Vauxhall's Q.C. said that there was still some land to the north of the site which might suit the organisers.

Representatives of Kelvinator had said that there was no need to alleviate a local unemployment problem. The Q.C. for Vauxhall Motors pointed out that the Kelvinators were claiming that the whole project was misconceived and that it was not necessary to alleviate any unemployment problem and that this view could not be accepted.

Vauxhall Motors did not belong to the Engineering Employers Federation which tried to achieve some control of wages during the 1950s and 60s. The motor industry was renowned for paying well above the average wage. This was a constant thorn in the sides of other large engineering companies in the Luton area such as Electrolux, Skefco Bearings, George Kent Instruments and Laporte Chemicals.[39]

Kelvinator had already seen a major competitor (Fisher Ludlow who were part of BMC and produced domestic appliances) open a factory at Kirkby and now three large motor manufacturers were also likely to come to their doorstep offering traditionally higher wages, good working conditions and steady employment. Their hidden agenda seems to have been that they did not want highly-paid auto workers from "The South" upsetting their local wage levels. They were also concerned that General Motors might want to transfer the production of refrigerators and tumble dryers from their Frigidaire Division in Hendon, London (another high-waged area) to Merseyside.

[39] Len Holden, *Vauxhall Motors and the Luton Economy 1960-2002*, Boydell Press

Wednesday's Hearing
Today Mr. T. P. Williams said that the company's existing factories were cramped and hemmed in and in planning a new factory they wished to avoid mistakes of the past.

He said that after consulting the Board of Trade they had looked at many sites on Merseyside but due to size, geology, closeness of housing and noise they were unacceptable. He agreed that only 175 acres would be needed initially, but further land would be acquired for future expansion. In response to the Agricultural Society's claim that they needed [for the Show] *between 102 and 130 acres, Mr. Williams said that they had looked at an aerial photo taken while the Cheshire Show was on and it was estimated that the show only occupied 65 acres.*

The Ministry Inspector asked Mr. Williams "Why did the company turn down a 900 acre site at Ince?" he replied that the site was marsh land subject to flooding and they had been told that solid land was 75 feet below ground level.

Mr. B. T. Whitam who lived at "The Limes" Hooton Park, described himself as an owner-occupier and said his home was within 50 yards of the proposed industrial area and if the development took place he would face a considerable financial loss. He withdrew his objection after receiving reassurances from Vauxhall Motors that they would buy his house at the present value without any depressed effect as a result of the development.

It seems that Mr. Whitam did not take up the Vauxhall offer. The author was told by a contemporary neighbour of Mr. Whitham that he continued to live in his house until the mid-1980s.

Thursday's Hearing.
Giving the view of the county council, Mr. Hetherington said the application was not being opposed in principle. The view was taken that because the new location on Merseyside was to be regarded as an urgent national need and it was on this assumption that the council had made its decision. Regardless of the result of the application someone would eventually have to face the future of Hooton Airfield for the County Council did not want it to become an eyesore.

The county council saw Vauxhall as the last major player in the motor industry to expand operations away from their traditional bases. BMC and Rootes had already committed to Bathgate and Linwood respectively in Scotland and the Ford Motor Company had announced plans only two months previously that it would expand to Halewood just outside Liverpool. So the pressure was on Cheshire County Council to grasp this employment opportunity especially as 20,000 overspill residents were coming from Liverpool.[40]

The Chronicle Saturday, June 25 1960

ROOM FOR VAUXHALLS, THE SHOW AND FARMER, INQUIRY TOLD

A 54- year old Hooton farmer would be faced with ruin and would suffer a monstrous injustice if Vauxhall Motors were granted outline planning permission.

On the concluding day of the inquiry, Mr. Price speaking on behalf of Mr. T. Unsworth of Stud Farm, Hooton said that his client began farming 163 acres at Hooton in 1939, but most of this was taken from him during the war. After the war he obtained a tenancy of the old Hooton Golf Course of 190 acres. He used the airfield to grow crops to feed his dairy and beef cattle. If the application went through this would mean ruin for Mr. Unsworth. Mr. Price went on to say that Vauxhalls would not be put out of business if they did not come to Hooton, but Mr. Unsworth could not go to any other place.

[40] Interview with Ray Bernie, 13 January 2009

Stud Farm was located on Rivacre Road to the west of Rivacre Valley on land eventually owned by Vauxhall Motors. It was last recorded in the 1967 electoral register and has now been demolished.[41]

> *The secretary of Cheshire County Federation of Ratepayers Associations said his organisation felt strongly that there was room at Hooton Park for Vauxhalls, the Cheshire Show and Mr. Unsworth. They felt that Vauxhalls should be satisfied with 350 acres at the most, rather than 400.*

It is interesting to speculate what qualified this body of ratepayers to advise the inquiry on the area of land that was needed for a commercial vehicle factory.

> *In his closing address Mr. Geoffrey Lawrence Q.C. (for Vauxhalls) said that whatever other concessions might be made, not less than 400 acres could be accepted. Vauxhalls had only come to Merseyside as a result of Government policy.*
>
> *At the beginning of the inquiry Mr. Lawrence had been under the impression that the Cheshire Show's difficulties were quite genuine and they had been forced to move from the Roodee to Hooton. After hearing the evidence it appeared that the show had moved because of a group of Agricultural Society people with expansionist and ambitious ideas.*
>
> *Mr. Lawrence said that Vauxhalls felt more sympathy for Mr. Unsworth than any other party at the inquiry.*
>
> *The inspector then closed the inquiry for the Minister's decision to be be announced later.*

On 29th July 1960 The Ministry of Housing and Local Government wrote to Vauxhall's solicitor saying that the Minister had considered the report of his Inspector Mr. M. B. Tetlow on the local inquiry, in which it was stated that he was not competent to assess whether 400 acres was a reasonable area for the scheme proposed. However, in the absence of any other information he was prepared to accept the figure as reasonable. He also doubted whether any other suitable site could be found although it was apparent that the local authorities and Vauxhall had considered alternatives. The development was within a borough in which some 20,000 overspill population were to be housed. He therefore recommended that permission should be granted but that it should be made personal to Vauxhall Motors. Other factors which influenced the Minister's decision were that none of the local authorities objected in principle. The Minister was impressed by the urgency of the economic factors and in the circumstances decided that it was right to grant permission subject to it being personal to Vauxhall Motors, the design of building be subject to local authority approval, the site should be screened and waste to be disposed of by means agreed with the local authority.[42]

These were the terms generally requested and recommended by the county council in their letter to the Minister on 3rd June.[43]

> **3rd August 1960, *The Times***
>
> ***Permission for Car Works Site***
> *Planning consent has been granted to Vauxhall Motors Ltd. for a commercial vehicle factory at Hooton Park, Wirral, Cheshire.*
>
> *This was granted after taking into account the urgency of economic factors: the high and persistent unemployment on Merseyside and the special need within the area for additional employment in an area which is to receive overspill* [residents] *on a large scale from other parts of the area.*
>
> *Conditions imposed covered the use of the site only by Vauxhall Motors, siting and design of the buildings, site screening, disposal of waste, and storage of minerals within the site.*
>
> *The Minister was pleased to note that Vauxhalls were prepared to co-operate, for the time being, with Cheshire Agricultural Society.*

[41] CRO Ref: CCR 1/879 [42] *op. cit.*, CRO Ref: LBE 7440/1/8 [43] CRO Ref: LBE 7440/1/5

Meanwhile – Across the Mersey

On 3rd February 1960 it was announced by Alderman Braddock during a meeting of Liverpool City Council that "I have in the last few minutes received confirmation that Ford Motor Company would expand their vehicle manufacturing operation to Merseyside."[44] He also said that there was also speculation that Standard Motors Co. may be here next; Triumph Herald bodies were already being built by Hall Engineering at Kirby.

Ford said in October 1961 that they had already received 9,000 applications for jobs at their new plant where it was expected that 8,000 jobs would be available in the first phase. During 1962 they set up a training assembly line in a hangar at Speke Airport using staff from Dagenham as instructors. When the plant opened in 1962 Ford wanted to get away from the multi-union approach at their Dagenham plants. They had reached agreement with the AEU and the NUGMW that there would be just the two unions recognised by the company. However, threats that the T&GWU would "black" Ford exports via Liverpool Docks gained them recognition as the third trade union.

It was initially agreed by the trades union that wages at Halewood would be 1s. 3d. per hour less than comparable jobs at Dagenham, but a series of overtime bans in 1963 soon gained Halewood parity with Dagenham.[45] Ford's building contractors were beset by labour relations problems which caused several months delay in starting production. In June 1962 Ford said they hoped to produce the first car by the end of the year, but it would be early March 1963 before the first Anglia De Luxe rolled off their assembly line. This car is now exhibited in Liverpool Museum.

The Plant Development

The Board of Trade had previously made it clear to all the big five motor manufacturers that if they wanted to expand in Britain, permission would only be granted in the designated "Industrial Development Areas" (IDA) so Vauxhall Motors set themselves certain criteria for the new site:

1. It should be built without unusual costs caused by extensive ground levelling, or piling.

2. As they would be recruiting large numbers of employees quickly they expected the new employees to travel from a larger area than employees at Luton/Dunstable where the workforce had built up gradually since the end of the Second World War and had consolidated in the immediate locale. Thus they would need access to a nearby motorway for employees and for interplant transport.

3. It should have plenty of space to be developed into a complete facility and allow for future expansion.

4. There should be no immediate housing areas that would be adversely affected by noise from the factory.

5. It needed both rail and dock or canal access.

6. It should not be close to similar industry to avoid competition for labour/wage levels.

It is presumed the company directors met occasionally with their competitors and exchanged views on their long-term plans and would undoubtedly have known that Ford and Standard were favouring the Liverpool side of the Mersey. Merseyside was the nearest IDA to the southern plants and it had large enough population to provide the necessary workforce. For these reasons it became the favoured area.

Looking at the five sites offered to the company at Bromborough, (south west of the power station), Hooton Park, Rossmore Road, Little Stanney (where the Cheshire Oaks Outlet Park is now located) and Ince Marshes, these sites all had the one advantage, that they were on the south side of the River Mersey and therefore not too close to Standard-Triumph and Ford who were rumoured to be looking at Speke and Halewood on the Liverpool (north) side of the Mersey.

In 1959 the only way across the Mersey was through the original and often congested Mersey Tunnel or across Runcorn Bridge. Although it is only 5 $^{1}/_{2}$ miles as the seagull flies it was at least a 21 mile journey and it was felt this was a sufficient barrier to discourage employees switching between companies.

The Ince site was about 900 acres and was therefore more than big enough but it was discovered that it would need piling to a depth of 75 feet in order to support the factory buildings and was ruled out for that reason alone. The Rossmore Road and Little Stanney sites were relatively small and were too near to residential areas. Bromborough would not have good enough trunk road access nor would any rail or canal facilities be likely.

This left Hooton Park and it quickly became the obvious choice as it met all the criteria set. It was relatively flat, the M53 was in the planning stage at the Ministry of Transport and would soon pass along the edge of the airfield, there were very few nearby houses and there was the potential to easily develop rail and canal connections.

Hooton Park had been requisitioned for a second time by the Air Ministry in 1939 for use as an airfield during the Second World War. By 1958 the Air Ministry no longer needed the land and had started negotiations with the Trustees of the Naylor Estates to return the land to them under the terms of the Crichel Down rules (that requisitioned land should initially be offered back to the previous owners). However, before these negotiations were completed the airfield came to the attention of Vauxhall Motors. Eventually a deal was brokered by the Air Minister, Lord Shackleton, to sell about a third of the airfield to the Naylor Estates and two thirds to Vauxhall Motors who also paid a sum of money to the Naylor Estates as compensation for their loss. A small part of the requisitioned land, which included part of Stud Farm, was at the same time earmarked for the Mid-Wirral motorway.[46]

Mr. Unsworth left Stud Farm in 1967 to become the tenant of another farm belonging to the Naylor Estates at nearby Great Sutton. Thus the person for whom Vauxhall had said during the course of the Inquiry that they had the most sympathy for retained his means of making a living and was not put out of business by Vauxhall's development.

The main source of information about the site development lies with the Ellesmere Port Borough Council building control records which have been deposited with Cheshire Record Office. These records include detailed planning applications and building control applications and the associated blueprints which have been catalogued and are available for public viewing.[47]

On 19th August 1960 a meeting initiated by the Ellesmere Port town clerk Ray Bernie for the mutual exchange of views and ideas was held between Vauxhall Motors, their architects, Howard Fairbairn and Partners and the structural engineers on one side and representatives of Ellesmere Port Borough Council, the county fire brigade, public health inspectors and building inspectors on the other side. Vauxhall presented their initial plans for the layout and structure of their buildings. For the architects Mr. Avann said that the initial problem was whether the local authority would insist on brick sheathing around steel structural members according to the bylaws 46 and 49, as this would considerably disadvantage Vauxhall in the utility of their buildings. He outlined the company's fire precautions and thought that there would be a case for relaxation of the bylaws.

Other matters raised, such as canteens, sanitary arrangements, welfare arrangements etc. met with general approval, but the council said that they would need more details before final approval could be given.[48]

This problem with bylaws 46 and 49 occurred repeatedly in the building control applications. The Deputy Borough Engineer had to arrange for a dispensation from the Ministry of Housing and Local Government on each occasion that plans were presented for building control approval.

[44] *Liverpool Daily Post*, 4 February 1960

[45] Huw Benyon, *Working for Ford*, Allen Lane Penguin Education, 1973

[46] Interview with Peter Poole, lately agent for The Trustees of the Naylor Estates, 15 May 2009

[47] CRO Ref: LBE 7440/1 for the applications and correspondance and LBE 7440/2 for drawings

[48] CRO Ref: LBE 7440/1/9

One design feature which gave considerable concern to the council surveyor's department was the use of overhead toilets in some factory areas. Vauxhall had used a design in their recent building at Dunstable where toilets and wash rooms were situated on mezzanine floors above the general work areas. This layout was already in use in their parts warehouse at Dunstable and the company wanted to adopt the same idea at Ellesmere Port. However, as this constituted a second storey the council bylaws stated that in such cases the steel supports had to be encased in fireproof brick or concrete for the general protection of life and limb. Planning officers went to view the Dunstable building in the belief that it was a factory where a similar bylaw had been relaxed. However, the building viewed was a parts warehouse where fewer people worked than would be the case in a manufacturing building and the Borough Engineer did not feel it was a comparable situation. Evidence was sought from the company of where the bylaws had been waived in a manufacturing building. Eventually the Ministry of Housing and Local Government did give permission to relax these two bylaws, but this had to be applied for in each subsequent building application.[49]

Vauxhall's building control applications for this initial building, EA block, and subsequent buildings were passed with only minor detail changes, these mainly being concerned with details affecting fire safety. Examination of the records for Vauxhall Motors applications at the Cheshire Record Office and comparison with other contemporary evidence suggests that in many cases requests for building inspections were submitted retrospectively after construction had commenced and in some cases only after the building work was complete. It has been suggested that this was not an uncommon practice on large construction projects because work would often start on one side of a building and steelwork would be erected before foundations were dug on the other side of the building. The council's building notification cards do not cater for this situation. In practice the building inspector was probably in daily contact with the contractor or visiting the site regularly to keep an eye on progress and only formalising the paperwork afterwards.

Short-term problems in the car market

While Vauxhall planned a long-term expansion in production capacity there were short-term sales problems as the market was in recession and on 7th October 1960 the company announced that they would have to give two weeks notice to 1,000 production workers. This would be restricted to Luton and Dunstable employees with less than six months service, the car factory would reduce operations to a four day/four night basis and as many workers as possible would be transferred to Bedford truck production at Dunstable as trucks were still selling well at home and in export markets.[50]

YEAR	EMPLOYEES	CAR OUTPUT	CV OUTPUT [51]
1959	26,251	157,365	88,720
1960	24,470	145,742	106,284
1961	23,584	90,549	95,839
1962	24,879	144,144	76,661
1963	30,843	164,287	84,789
1964	33,500	236,226	106,646
1965	33,022	220,807	112,630
1966	33,476	174,878	100,505

Table 1: Vauxhall total employees and production by year.[52]

[49] CRO Ref: LBE 7440/1/16

[50] *The Times* newspaper, various dates, October 1960

[51] The commercial vehicle output includes both trucks built at Dunstable and CA vans produced at Luton

[52] Sources: *Vauxhall Facts and Figures 1966* and *Vauxhall Information Handbook 1978*

In 1961 car output was down some 38% on the 1960 figure whilst commercial vehicle output was only down by 10%. When car production picked up in 1962, commercial vehicle production continued to slide and workers were transferred back to Luton. The employment levels increased from 1963 as Ellesmere Port recruited new labour, this resulted in the increase in car output of about 100,000 between 1962 and 1964/5 as forecast in the annual report of 1959. It was during the autumn of 1960 that trades union officials realised that, if van and truck production was moved 160 miles north to Ellesmere Port, the ability to switch workers between car and commercial vehicle production would not be practical, and in any future recessions there could only be one outcome, more lay-offs. Dunstable Council was also concerned that this inflexibility could adversely affect the local economy.

The company changed their plans in June 1961. After considerable lobbying by the council, local papers and trades union officials, the company said that instead of Ellesmere Port being a centre for commercial vehicle production it would now be a feeder plant for Luton and Dunstable. Instead of truck and van manufacturing, it would make mechanical components. They said that "to move truck assembly would cause too much disruption to production".[53] This was a smokescreen. The company would not admit to bowing to outside pressures. The change of plan did not negate the outline planning application, as it had always related to "Motor Vehicle" manufacture and did not refer specifically to commercial vehicles, cars or component manufacture.

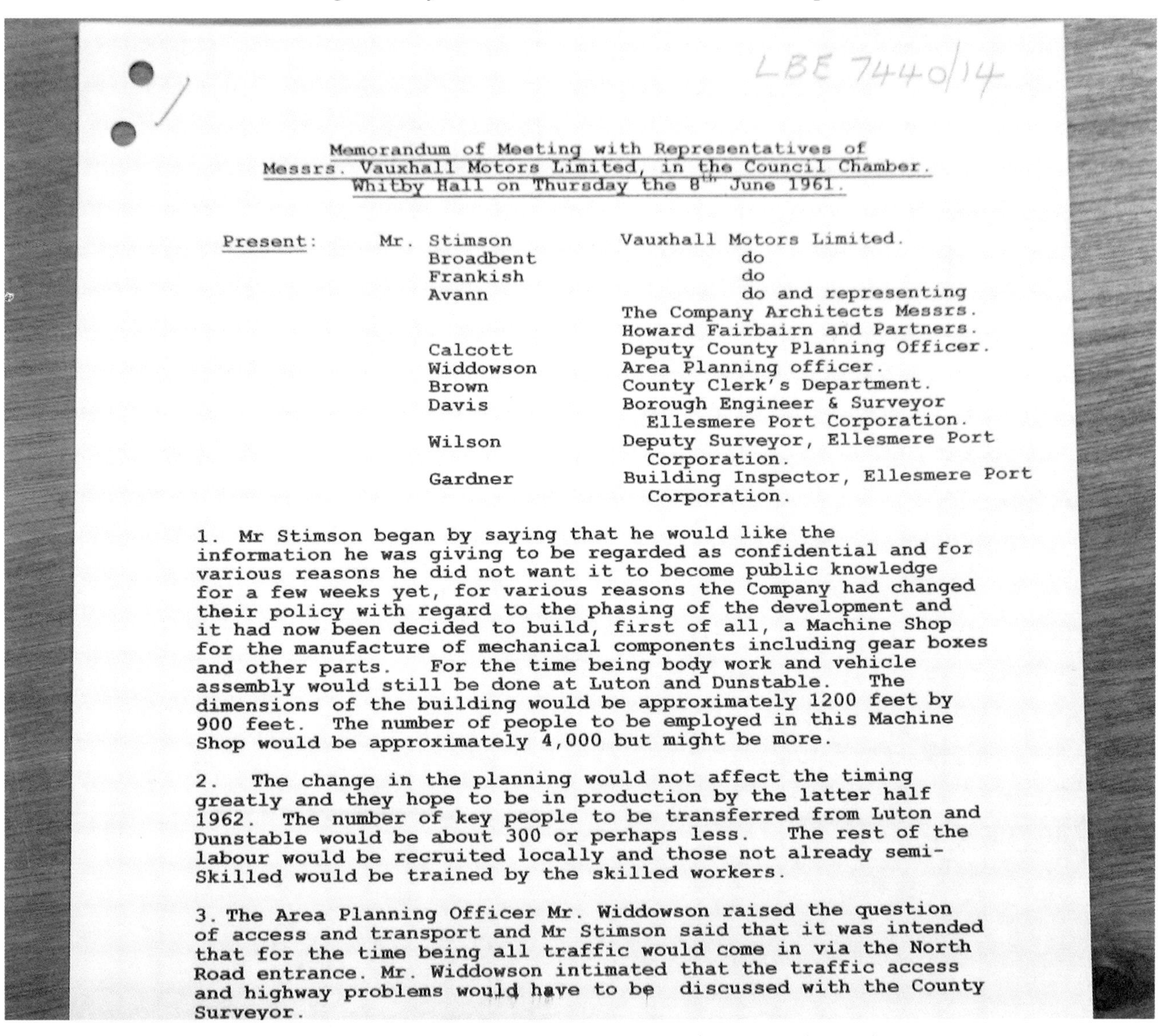

LBE 7440/14

Memorandum of Meeting with Representatives of
Messrs. Vauxhall Motors Limited, in the Council Chamber.
Whitby Hall on Thursday the 8th June 1961.

Present: Mr. Stimson — Vauxhall Motors Limited.
Broadbent — do
Frankish — do
Avann — do and representing The Company Architects Messrs. Howard Fairbairn and Partners.
Calcott — Deputy County Planning Officer.
Widdowson — Area Planning officer.
Brown — County Clerk's Department.
Davis — Borough Engineer & Surveyor Ellesmere Port Corporation.
Wilson — Deputy Surveyor, Ellesmere Port Corporation.
Gardner — Building Inspector, Ellesmere Port Corporation.

1. Mr Stimson began by saying that he would like the information he was giving to be regarded as confidential and for various reasons he did not want it to become public knowledge for a few weeks yet, for various reasons the Company had changed their policy with regard to the phasing of the development and it had now been decided to build, first of all, a Machine Shop for the manufacture of mechanical components including gear boxes and other parts. For the time being body work and vehicle assembly would still be done at Luton and Dunstable. The dimensions of the building would be approximately 1200 feet by 900 feet. The number of people to be employed in this Machine Shop would be approximately 4,000 but might be more.

2. The change in the planning would not affect the timing greatly and they hope to be in production by the latter half 1962. The number of key people to be transferred from Luton and Dunstable would be about 300 or perhaps less. The rest of the labour would be recruited locally and those not already semi-Skilled would be trained by the skilled workers.

3. The Area Planning Officer Mr. Widdowson raised the question of access and transport and Mr Stimson said that it was intended that for the time being all traffic would come in via the North Road entrance. Mr. Widdowson intimated that the traffic access and highway problems would have to be discussed with the County Surveyor.

Figure 44: Vauxhall explain to the council that their plans have changed.

[53] CRO Ref: LBE 7440/1/14

Choosing a name for the plant

Some confusion seems to have existed regarding the name of the plant. From documents in the Ellesmere Port Borough Council files, it seems that the plant was initially to have been called the "Hooton Park Plant", with the first building being designated as HA block. But for several reasons this was soon changed to the "Ellesmere Port Plant" with the first building being designated EA Block.

1. "Hooton" could easily be mistaken for "Luton" over the phone. At Luton planning was underway for a major new vehicle design headquarters, to be known as "AJ" block, again easily mistaken for "HA". These phonetic similarities were foreseen as a cause for many errors "twixt ear and lip".

2. Also the next major car project had already been designated "HA Viva".

Figure 45: 1963 Vauxhall Viva, model HA, price £527 including purchase tax.

It is common Vauxhall parlance to refer to all buildings by their block designation letters rather than a descriptive title and cars were always referred to by their model letters, for example HA and HB for Viva and F-type for Victor.

At this point you may have spotted that I usually refer to the company by its shortened name as "Vauxhall" in the singular whilst the book title adopts the plural version. The title purely reflects the Merseyside dialect "Vauxhalls" when talking about both the company and the plant. Southerners generally use the singular version. Also residents of Ellesmere Port often speak of their town as "The Port".

5th October 1961, *Ellesmere Port Pioneer*

Visit to Luton and Dunstable.
The Mayor and Trades Council today visited Vauxhall factories to view the van production line at Luton and the truck plant at Dunstable. This gave them the opportunity to see what problems would be involved when Vauxhall extend their activities to Ellesmere Port.

Why this group chose to visit the commercial vehicle productions facilities remains a mystery, as the plan had been changed in favour of a car plant some four months previously.

Planning and building work

Figure 46: Joe Davies.

Howard Fairbairn and Partners Ltd. had already been appointed as architects for the project by Vauxhall Motors. Once outline planning permission for the factory had been given by the Ministry of Housing and Local Government, the firm of Sir Alfred McAlpine (Northern) Ltd. was appointed as the main contractor for the whole site and the buildings. This contract would be worth £18.1m to McAlpines during the construction period, 1961 to 1970.

Vauxhall took ownership of the site on 25th July 1961[54] and construction began in the following month. The structural steel work for EA Block started on 16th October 1961.

Mr. Joe Davies, as a 21 year old tractor driver was working for his father at Park Farm and remembers that when McAlpines came to peg out the site, before any construction could start they found that the grass was too long to see the pegs. He thinks it was one of the McAlpine brothers who gave him ten shillings to mow strips in the grass so that the site could be accurately pegged out. He can, with some justification, claim the honour of being the first person to do any physical work on the plant.[55]

The site, having been an airfield, was of course fairly level and flat, one factor which had led to the selection of Hooton Park for the factory. At Luton, where later factory building work was done on very sloping ground, changes in level led to some peculiar building layouts; one building had three floors, each with ground level entrances.

Building work was to commence at Hooton (sorry, Ellesmere Port) with EA Block. This part of the site was occupied by some redundant Second World War sheds which had been used for barrage balloon storage and a scrap engine dump. These were quickly demolished. The ground level rises from about 50 feet above sea level at North Road to about 90 feet near the small plateau on which Hooton Hall once stood. This plateau rose sharply about another ten feet.[56]

The design of two of the main buildings, EC and ED Blocks, took advantage of the general rise in ground level by incorporating basements on their north east facing sides. This was particularly beneficial in ED Block. The sheet metal presses, being massive structures, were operated at the ground floor level while much of the lower mechanism "hung" into the basement. Sheet metal trimmings from the pressings fell by gravity onto conveyors leading to the scrap balers.

Figure 47: Tom Williams (Vauxhall's Project Manager who was later the first Plant Manager) cuts the first sod in August 1961 supervised by Ivor Avann (architect) of Howard Fairbairn and Partners Ltd. Also in the picture are (back row right to left): Dennis West (Company Treasurer), Eric Fountain (Chief Factory Layout Engineer) and Roland Broadbent (Works Engineer). Far left in the back row is Bill Smith (Works Engineering Manager). Below Eric Fountain is Frank Halesworth (Buildings Engineer). Seated, waving his hat is Ted Suggett (Quantity Surveyor) of G. D.Walford & Partners. Front row third from left is Charles Potter (Clerk of Works) of Howard Fairbairn and Partners Ltd. Second from right is Charles "Stimmy" Stimson (Vauxhall's Assistant Company Secretary). Apologies to other people who have not been identified.

[54] CRO Ref: LBE 7440/1/14 [55] Interview with Jane Davies, 3 July 2008 [56] CRO Ref: LBE 7440/2/37 & 40

Timescale for other buildings

EA Block was the first main production building to be built and would be used for the manufacture of engines, gear boxes and axles for the HA Viva, and the F-type Victor car and gear boxes for Bedford vans and trucks.

While EA Block was under construction, work was progressing on the boiler house. This would generate steam for factory heating and compressed air which was used for many power tools. A telephone exchange, a fire station, an oil and paint store, security lodges and the trade effluent plant were all part of the first phase which was completed by the end of 1963.

Early during the installation of equipment it was found that the main engine assembly conveyor had pedestals three feet high and returned on a loop under the main conveyor. This return loop was in a trough about two metres deep. The trough kept filling with water at the times of particularly high tides in the nearby River Mersey. Investigation revealed that, although the factory floor was about 55 feet above mean sea level, the local water table rose and fell with the tides. Lining the trough with an impermeable skin cured the problem.

At the end of 1962 a foundation stone was laid adjacent to the main reception entrance to EA block by Dr. Charles Hill, M.P., for Luton and previously BBC's "Radio Doctor".

Figure 48: The first steel work for EA Block being erected in November 1961.

EC Block

EC Block was the second major factory building in the project and construction began on 19th June 1963. This building was unusual because its size and purpose were subject to many changes during the planning stages and it was constructed in three phases. There is one plan showing it having an area of over 3 $^{1}/_{4}$ million square feet.[57] At that size EC Block would have been three times the size of EA Block and would have covered just over 70 acres (Heathrow Terminal 5 has a ground floor area of a mere $^{3}/_{4}$m. square feet). However, a building of just 871,440 square feet was eventually constructed. The first phase (1964) had a pilot car assembly line. This was used for training production operators in car assembly operations. Later, when ED Block become operational, this pilot assembly line was scrapped. Facilities for radiator and petrol tank manufacture, tool rooms, miscellaneous machining and CKD packaging were installed. The basement was used for the machining of non-current spare parts.

[57] CRO Ref: LBE 7440/2/31

Vauxhall Firenza SL, a 2 door coupe version of the HC Viva, July 1971

yours vivaciously

THE '65 VAUXHALL Viva

Did Woman have the first as well as the last word about the design of this spacious, vivacious Vauxhall? Take size . . . big-car good looks are cleverly combined with small-car parkability. Viva is family size inside; family-safe too and certainly has a family-holiday-size boot. Then take performance—Viva acceleration makes motor-minded husbands whistle and then rave about "0 to 50 in 13·3 seconds": but Viva has that easy-going top-gear stride that keeps your hand off the gear lever . . . and also cuts down your calls at the petrol pump. Not that you mind changing gear with this sweet, all-synchro gearbox; it just makes you an expert.

But the smart thing to do is to try a Viva. Run a shrewd shopper's eye right over it: study the colour range and the new two-tone, de-luxe upholstery; but do try the ride Look at the price tag and do a bit of figuring: you'll agree that a Viva is quite remarkable value.

Make your Viva test-drive a family occasion. See how it's been built around you . . . a car that fits in with your family . . . not a car that the family has to fit into . . . a very practical car that offers you more of everything that matters.

Parking a Viva needs less effort than you'd thought possible. Pedal pressures are very light indeed. You can see all four "wing-tips" from the driving seat and your Viva can turn right round inside a 29-ft. circle.

Headroom for a high-style hair-do: hatroom for a grand occasion. Easiest possible access to the rear seat (even in a pencil-slim skirt), so getting out can always be graceful and elegant.

One good reason why the Viva is designed the way it is, is to give you this jumbo-size boot. Everything for the whole family goes in when you go off on holiday. And what a boon for a weekend shopping spree!

Every Viva receives a complete underbody seal, including wings and wheel arches, at no extra charge. Bodywork, besides being chemically rust-proofed inside as well as outside, is deep-dipped into special proofing paint for inside/out protection. Primer coats, colour coats and the final deep gloss coats of "Magic Mirror" acrylic lacquer are baked on. Viva value is lasting value!

CHOICE OF EIGHT FASHION COLOURS

Exterior	Upholstery (Vynide)
*Grecian White**	*Red*
Black	*Red*
*Storm Grey**	*Red*
*Meteor Blue**	*Blue†*
Pacific Blue	*Blue*
Jade Green	*Beige*
Cavalry Fawn	*Beige*
*Calypso Red**	*Beige*

**Colour schemes for the standard Viva*
†Beige upholstery for the standard Viva

Easiest possible car to clean. Just wash, rinse and wipe to restore showroom shine. Polishing is out for Viva families. Aluminium grille can't rust. Hubs and bumpers have new tripled resistance to corrosion.

This is the new Viva de Luxe rear seat—in smart, soft but tough and easy-to-clean Vynide. Note the wrap-round effect for comfort in big-car style. Spacious: yes, like the well-shaped, well-padded front seats—take another look at them on the front cover of this brochure.

It is easy to drive well behind this wheel . . . with the lightest, most precise steering and pedals which take much less effort to push.

Positively no excuse for fiddling under this bonnet! But boys will be boys, so let them admire what Dad calls "the clean, accessible layout of the engine compartment". Anyway, routine checks don't dirty shirt-cuffs any more. And, believe it or not, he'll only have the "fun" of getting underneath with his grease gun once every 2½ years!

Viva

TWO VIVA MODELS TO CHOOSE FROM

Viva de Luxe (illustrated throughout this brochure) and the standard Viva. They are identical in size and mechanical specification but the Viva de Luxe has extra sound insulation, padded dash, edged carpets (instead of rubber mats), heater, special seating, two-colour door trim, doorpull armrests, hinged rear quarter lights, passenger's sun visor, screenwash, bright interior fittings, bright drip channels on roof, bright surrounds to windscreen and rear window, bright body side moulding, mat on boot floor. Both models have headlamp flasher, seat belt anchorages and are underbody sealed.

Viva: £528. 7. 11. inc. P.T.
New Viva de Luxe: £573. 2. 1. inc. P.T.

VAUXHALL MOTORS LTD.
LUTON, ENGLAND

Thanks a million!

Thanks to YOU, for helping us make a million Vivas. Thanks to you in Manchester and Montreal, to you in Rochdale and Rotterdam, you in Birmingham and Basle, Vancouver and Vienna, Cardiff and Calcutta, Margate and Milan, Barmouth and Brussels. Thanks, all of you; without you we would not have built our millionth Viva. We did that on July 20, at Luton. A happy milestone. We're happy to have had so much help from so many of you to reach that milestone; cars without customers would have got us nowhere. Viva first appeared in September '63. We built 309,538 of those HA models. In September 1966 came Viva 2, the HB series; by October last year we had built 556,752 of those – thanks to your support again, of course. Then last October came the current HC Viva, and we built our first 100,000 of those in only seven months – our fastest 100,000 ever. As you gathered from our first few lines above, you Viva people are pretty well scattered over the globe. Exports of HA and HB Vivas totalled over 337,000; current HC exports are over 46,000. So you're in good – and varied – company, all of you. On July 20 there were salutes all round, especially to the men who built and sold those million Vivas. This is a salute to you, the customers who bought them. Turn over for a few flashbacks, 1963–1971. And once again, thanks a million. Stay with us, won't you; there are a few more milestones we'd like to reach yet.

29

July 20 1971, Vauxhall mark the 1,000,000th Viva being produced, by chance this magic number was a 2.3 litre car. For a short period these larger engined Magnum/Vivas were produced at Luton.

Figure 49: Preparing to lay the "Foundation Stone" 12th November 1962, left to right: Mr. (later Sir.) William Swallow (Chairman and M.D.), The Rt. Hon. Selwyn Lloyd (M.P. for Wirral), Dr. Charles Hill (M.P. for Luton), Sir Reginald Pearson (Deputy Chairman) and Thomas P. Williams (Plant Manger (USA)).

ED Block

In 1965 work commenced on ED Block, the main press shop and car assembly building. This can be seen in Figure 51 which shows the press shop under construction in the summer of 1965. The concrete pillars of the basement have been built up to the main ground floor level and the steel work for the body shop and paint shop can just be seen on the left-hand side.

In December 1965 35,000 cubic yards of concrete were poured. Before this, however, half a million cubic yards of earth had been shifted from site, half a million bricks had been laid with four million in reserve. Over 400 piles had to be driven into the body shop area as some previously filled-in ponds were discovered when earth moving was being done. This had not been revealed during soil surveys and the ponds had been filled with top soil and vegetable matter. At the peak 1,500 contractors' men were on site. There had been no major labour disputes on the site, but there was a nationwide shortage of cement. Frank Halesworth, Vauxhall's Chief Building Engineer reported that he "...doubted whether any other project in the country was importing ready mixed concrete in the quantities we are". ED Block was completed and operational for the introduction of model HB Viva by August 1966.

Figure 50: Looking south east early in 1963 with EA Block on the left with the mud and equipment typical of a large scale construction project.

ED Block was to be the last main production building on the site built by Vauxhall Motors, however, EF Canteen and EG Office Block were started in the spring of 1966 with George Wimpey Ltd. as the main contractor. This brought about the "interesting" situation of having two major civil engineering contractors on the same site.

After the canteen and offices were finished no further major developments to the plant were made. Internal alterations to buildings to accommodate changes in production methods and processes would be a frequent occurrence in succeeding years. Subsequently some land was sold for suppliers to build warehouses and sub-assembly facilities to meet the lean, "just in time" philosophy of manufacturing.

Figure 51: ED Block, summer 1965.

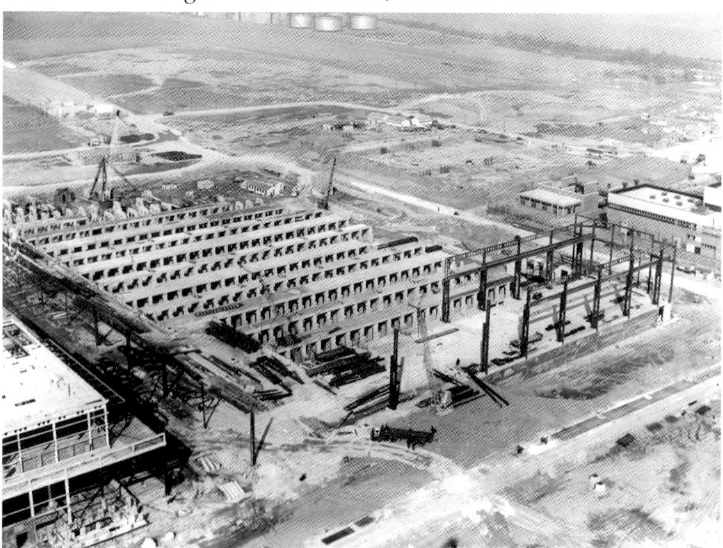

Figure 52: A 1966 HB Viva 2 door saloon.

In January 1966 the borough council wrote to Vauxhall suggesting that the area being developed might exceed that requested and granted for the Industrial Development Certificate in February 1960. They asked for a breakdown of the floor areas. This was supplied by Vauxhall in September 1966 (see table 2), and shows that the original area had been exceeded. However, no further action could realistically be taken by the council as they had approved each individual application.

In September 1967 the Cheshire County Council Area Planning Officer wrote to Vauxhall's managing director pointing out that "...*one of the original conditions for permission to build laid down by the Ministry of Housing and Local Government was that the site should be screened as may be agreed with the local authority and so far as he was aware no proposals have yet been submitted*...etc".

A proposal by consultants for landscaping was submitted in November 1967 to Cheshire County Council. This detailed the species and location of trees on the land between ED Block employee car park and the new motorway to break up the industrial skyline. The council generally agreed with the proposals and only suggested a change in one species of tree to "*something less exotic*".[58]

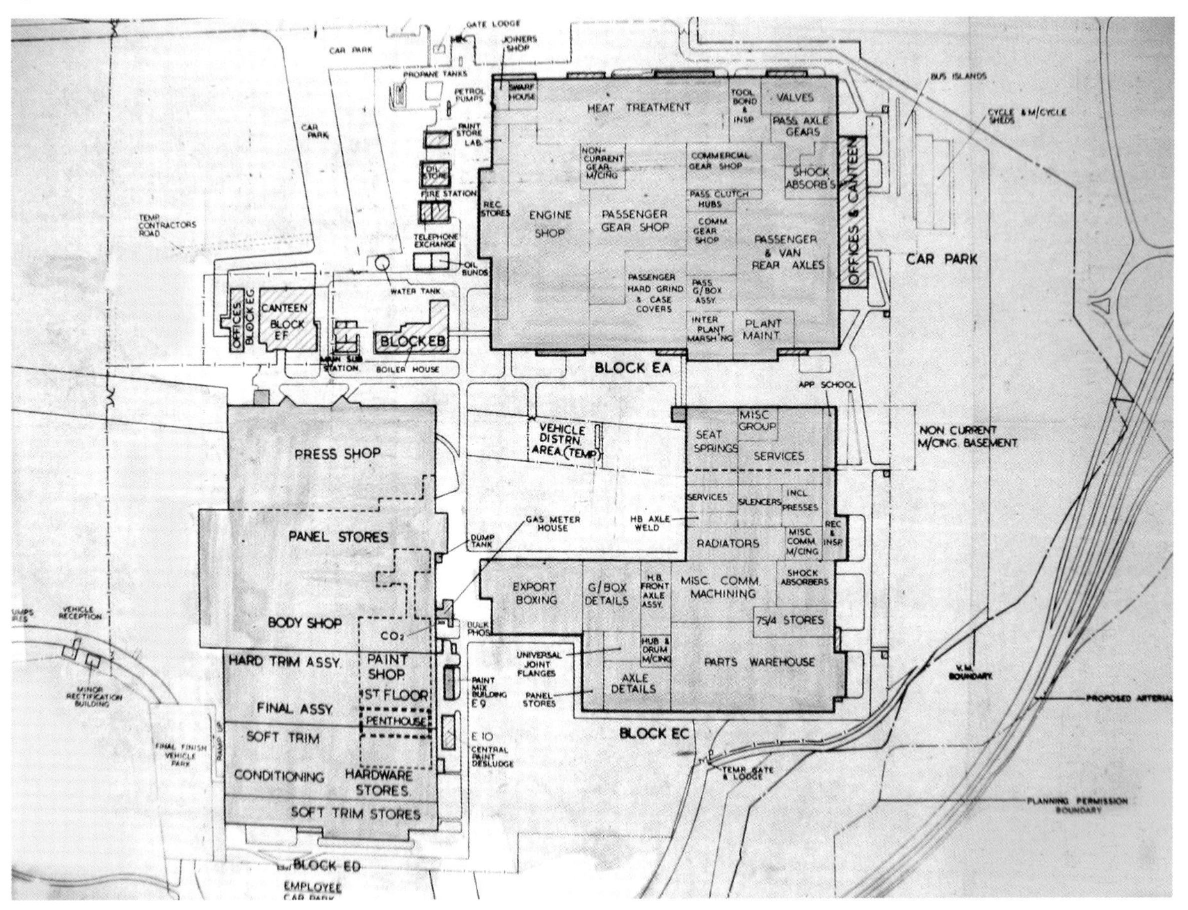

Figure 53: Plant layout as finished in 1970.

Location	Floor Area - Square Feet	
	Industrial Buildings	**Offices, Canteen and Separate Service Buildings**
Block 'EA'	1,063,000	103,330
Block 'EC'	865,000	6,440
Block 'ED'	1,265,461	8,864
Miscellaneous Buildings	4,628	252,981
Totals	3,198,089	371,615
Miscellaneous Buildings areas include some original airfield buildings which all lie on the Bebington side of the Borough Boundary.		

Table 2: The floor area of the major buildings in 1966.

[58] CRO Ref: LBE 7440/1/48 & LBE 7440/2/20

Trade Effluent Treatment [59]

Early during the plant development a trade effluent treatment plant was built between North Road and the ship canal as every large car plant produces a variety of waste liquids including:

Rainwater surface run off
Soluble and neat cutting oils
Water from component washing machines.
Acid and alkali waste from paint processes
Kitchen waste fluids
Hand washing waste water
Human foul waste.

Originally all surface water from the old Hooton Airfield was fed via land drains into a pond known as "The Lagoon" which discharged directly into the Manchester Ship Canal. This system has largely fallen into disuse because over the years much of the original grassland had been covered with roads and buildings making the land drains redundant.

Rain falling on the plant is now taken by a new drainage system into a stream, "The Ravine" which has been dammed to control the outlet rate in surge periods by providing a balancing pond and smoothing the rate of flow into the ship canal via a weir. An interceptor is used to remove oil and grit from the flow and the oil is skimmed off and re-refined. This system prevents any oil contamination of the canal water. The final outflow is slowed down and broken up by a number of large rocks placed immediately before the water enters the ship canal and reduces any side thrust on slow-moving ships.

When the plant opened in 1962 a large amount of both neat and soluble cutting oil was used in the various metal machining processes. This waste oil was collected, together with liquids from component washing machines and pumped to the effluent plant where the oil was split from the water and then sent away for re-refining. Sometimes the oil still contained too much water and that also was sent away for further treatment and re-refining. Acid and alkali effluents were chemically treated and neutralised, the sludge removed and sent to landfill sites. Water from these processes was discharged into the ship canal. Most of the time the discharge looked like fresh water and it was said on many occasions that Vauxhall Motors discharges improved the quality of the canal water. Hand washing water, human foul waste and kitchen water were all discharged into the local authority foul sewer.

Plans of the Trade Effluent Plant can be seen under CRO Ref: LBE 7440/2/2 and 6.

One modification made to the trade effluent plant after the start of full-scale production involved cooling water for the main air compressors in the boiler house. The same water was also circulated through all the spot welding guns in the body shop, so was overheating and reducing the compressor's efficiency. A solution was found by George Lynas, the works engineer who arranged to build a fountain in the Ravine to which this (hot) cooling water was pumped, cooled in the fresh air and then returned to the compressors.

Figure 54: Pouring concrete for the outfall to the ship canal.

[59] *Trade Effluent* section written by John Morton

Start of Production

12th November 1962 was an important day for Ellesmere Port as the first locally recruited employees started their careers at Vauxhall Motors Ltd. One in particular, David Dryer, served 37 years in the accounts department, and then in his retirement carried on as a part-time factory guide.

> **15th November 1962, *Ellesmere Port Pioneer***
>
> ***Vauxhall Factory is in Production***
> *On Monday* [12th] *machines which had arrived only a few days before began turning out gears, steering and engine parts for cars and trucks which are non-current spares when 40 new production workers started. They were introduced to their jobs by the first key men transferred from Luton. In addition to the production operators 35 other local employees – electricians, material handlers, store men etc. also started work on Monday.*

Transfer of machine tools to Ellesmere Port

Non-current spares (for previous model cars) were the first parts to be produced at the plant. This did not involve proving any new machine tools, nor was the production of finished vehicles dependent on the quantity of parts produced. This was an ideal way to introduce local labour to motor industry production methods, albeit on a small scale. Non-current spares were manufactured on a batch production basis on relatively simple and general purpose machine tools which had been transferred from Dunstable during the previous two weeks.

Recruitment of local employees

On 12th November 1962 the first local employees, most of whom were new to the motor industry and would not be familiar with mass production methods, started work with the manufacture of non-current spares. A few of the new starters however did come from Girling Ltd. at Bromborough who were one of the two main suppliers of automotive brakes. These ex-Girling employees had a background in the motor industry and it was from this nucleus that the first machine tool setters were selected. Vic Honey, who can be seen in figure 55, was the first local employee to be appointed as a setter, then a foreman and finally as general foreman in the commercial gear shop.

Figure 55: New employees are introduced to their foreman, Barry Marshal and setter, Sid Smith, by area manager Jim Spence in November 1962. At the back of this group is Vic Honey.

Alongside the new local workers were experienced staff, who had transferred, either on a temporary basis or as permanent key staff, from Luton and Dunstable. During 1963 there were about 240 temporary transferees and 80 key people who soon moved home from the Luton area. They were all involved in the initial training or staff and supervisory functions.

Transfer of current production facilities to Ellesmere Port

During January 1963 the production of gearboxes for the Vauxhall Victor was moved from Luton. The first element to be moved was the gearbox assembly line. Stockpiles of finished gearboxes had been built up at Luton and these were used to keep the car assembly line working during the two weeks delay while the gearbox assembly line was moved. Once this assembly operation was running at Ellesmere Port, the gearbox component machine tools were moved north. During this move previously stockpiled components were used to feed the gearbox assembly line. A similar process was involved when the Victor rear axle assembly line was moved from Luton to Ellesmere Port during February 1963.

During the winter of 1962-3 some 4,500 machine tools and pieces of equipment were moved from the Luton and Dunstable plants to Ellesmere Port. As an example of the speed of the whole operation, a Snyder transfer machine with 40 individual stations for the gearbox cases was installed, fully aligned, rewired and producing components within 21 days.

Figure 56: The rear axle assembly line, 1st March 1963.

At the same time as existing production facilities of Victor gearboxes and rear axles were being transferred and set up, a totally new car, Viva (HA), was in the last stages of pilot build. This car was to be assembled at Luton with full-scale production starting in the autumn of 1963.[60] However, the mechanical units i.e. engines, gearboxes and axles would be produced entirely at Ellesmere Port and the first complete set of Viva mechanical units was built on 23rd April 1963.

[60] The Viva's retail price was £436 excluding purchase tax

Figure 57: The first set of Viva mechanical produced on 23rd April 1963, left to right: Aubrey Allison (engineering), Jim Govan (manager – inspection department), Jim Spence (area manager – axles), George Barker (area manager – gear boxes), Reg Burson (general foreman – gear boxes), Frank Ash (production manager), Tony Potter (assistant production manager), Arthur Kaye (area manager – engines), Tom Williams (plant manager), Ted Gilbert (supply manager), Tom Skidmore (manager – methods engineering), Bill Barker (superintendent – tool room), Charlie Hyde (manager – production engineering).

This was a testing time for the venture with a new product, produced on new machine tools in a new factory and by a new workforce. In addition to the main production at Luton from 1st June 1964 the HA Viva car was also partly built at Ellesmere Port with just the final assembly being done on a pilot line in EC Block. The previously painted bodies were transported by rail from Luton. This line was used to train assembly operators in readiness for the start of full-scale car production when ED block was completed. A production rate of just ten cars per hour was possible, although it was only rarely that eight cars per hour was exceeded.

Figure 58: Charlie Dunn (area manager) stands beside the first Ellesmere Port assembled Viva as it rolls off the pilot assembly conveyor in EC Block.

This first conveyor was unusual in that it was not a continuously slow moving conveyor; a bell rang and everyone had to stand back while the cars moved forward one station at regular six minute intervals.

HA Viva built in Luton had been launched to market in September. A contemporary advert, figure 61, shows the car which was offered in two versions, Saloon at £527.7.11 including purchase tax and Deluxe saloon at £566.1.3 including tax. The only difference between the two models was the addition of a heater and windscreen wash on the Deluxe model.

Figure 59: Pilot line, soft trim section.

Figure 60: Pilot line, the marriage conveyor.

Compare the Viva with any car in its class. World-wide tests have proved its superiority in all these points: speed, handling, steering, roominess, suspension, and all-round finish. But drive one, find out for yourself!

Real comfort for four
S-T-R-E-T-C-H-I-N-G space for four big people. In all, more interior room *than in any comparable car.* Wide doors for easy access. Real 'millionaire' comfort in the specially sprung and padded seats.

Big car performance
Undoubtedly fastest in its class. Top speed 80 plus. Over 70 in third. Beats comparable cars in all speed ranges. Four-speed all-synchromesh gearbox. High performance engine. Optional disc brakes.

Roll Control Suspension
Viva's suspension will amaze you. You get 'millionaire' smoothness even on very rough surfaces. Sports car cornering with none of the pitch-and-toss some light cars have. Unique 'roll control'.

Best Steering and Handling
Carefully designed front-engine layout gives perfect handling and very easy access for service. Viva's steering effort is *lightest in its class.* Turning circle 29 ft., ideal for tight parking.

Carefully Designed Controls
All controls easy to reach. Headlamp dipper, flasher, horn and direction indicators controlled by one lever on steering column. Short gear-lever and handbrake fall perfectly to hand. Maximum all-round vision.

Enormous Boot
Viva first again on luggage room. Space for 10½ cu. ft. of hard suitcases. Lots more for soft baggage. The Viva's finish is meticulous. 5 gallons of paint protect the exterior. Complete underbody seal is standard.

VAUXHALL
VIVA

Vauxhall Viva, 4 cylinders, 1057 cc.
Saloon £527.7.11 (£436+£91.7.11 p.t.)
De Luxe Saloon with heater, Screenclean, etc., £566.1.3 (£468+£98.1.3 p.t.)

DESIGNED FOR OUT AND OUT RELIABILITY

See the Viva at your Vauxhall dealer's now!

Figure 61: HA Viva advert from October 1963.

3

Figure 62: Pilot Line just after the marriage conveyor.

In 1966 the HB Viva started its production life in ED Block. Production capacity was then 100,000 vehicles per year. Ellesmere Port then became the only plant producing the engines, gear boxes and axles and then assembling the Viva. This was in addition to all the components manufactured and taken to Luton and Dunstable for assembly into cars, vans and trucks built at the southern plants.

Employment at the plant increased steadily until 1968 and then fluctuated due to car market conditions. At one stage nearly 12,000 people were employed at the Ellesmere Port plant.

Date	No. Employees
31 Dec 1962	389
31 Dec 1963	4140
31 Dec 1964	5480
31 Dec1965	6043
31 Dec 1966	8856
31 Dec 1967	10859
31 Dec 1968	11752
31 Dec 1969	10286
31 Dec 1970	11992
31 Dec 1971	11701
31 Oct 1972	11440

Inter-plant transport operations [61]

Once the decision had been taken in 1960 that Ellesmere Port would be the site for a multi-million pound expansion for Vauxhall Motors to cope with the rising world demand for its products, there began a period of transport planning. The new transport system needed to serve the existing plants at Luton and Dunstable. Initially Ellesmere Port was to be a feeder plant producing a host of mechanical components and sub-assemblies for the final assembly lines 168 miles away in the south. The plant activities would have to be very tightly linked, the components produced at Ellesmere Port having to be fed with unbroken regularity to the southern plants, in effect a 168 mile conveyor belt!

This was the first time that Vauxhall Motors had been involved in a regular, large scale trunk haulage operation. The system needed to meet four main requirements:

1. The most economic way of hauling 48 return loads per day between north and south; this figure took account of the likely continued growth of production and the growth of the Ellesmere Port plant itself.

2. To ensure a constant flow of materials, so that production at Luton and Dunstable was not disrupted in any way.

3. To enable the staff in charge of actual production, at all three factories, to have precise control of vital material.

4. To give drivers normal working conditions without having to sleep away from home.

[61] John Peake, *Development of Hooton Park*, unplublished Ms. document, *circa* 1964, CRO Ref: 3888

Figure 63: A convoy of "Artics" makes the early morning run to Baginton.

Three methods of transport were considered: by an outside road haulier, by rail or by Vauxhall's own transport fleet. Using the company's own fleet won hands down, not only on the basis of cost, but there were other advantages. A company owned fleet of Bedford trucks would allow much tighter control of material movement and promised greater flexibility to cope with fluctuations in production needs. It also made sense for Britain's largest producer of trucks to use its own vehicles with the added prestige of a smart well-kept fleet bearing the company's name. An additional bonus of having, for the first time, their own fleet of long-haul trucks was that vehicle engineers had their own extended test facility for prototype or modified parts.

This was obviously a job for "artics" (tractor units with articulated trailers). Rigid trucks would not allow the flexibility of operation that was vital to get the optimum use of the vehicles. To allow drivers to sleep regularly at home, an intermediate staging post was a "must". The point chosen for this was at Baginton, just off the A45 south of Coventry. This was 65 miles from Luton/Dunstable and 103 miles from Ellesmere Port. The system worked well from the outset. The Ellesmere Port drivers, with no motorway miles, completed one return journey each day. Drivers based in the south were able to complete two return trips every day because for their 65 mile journey, 53 miles were on the new M1/M45 motorway. This meant that two drivers were required at Ellesmere Port for every one based in the south and this went towards meeting the original Board of Trade aim of reducing unemployment on Merseyside. The trucks themselves were Bedford TK tractor units using Bedford 330 cubic inches diesel engines, with an artic semi-trailer. They had a gross combination weight of 17 ton 3 $^{3}/_{4}$ cwt.

The artic semi-trailers were 26 feet long and within the legal maximum combined length of 35 feet for tractor and trailer. The ideal body, it was decided, would be fully enclosed with two alloy sliding doors on each side. This would allow the side-loading of stillages and pallets by fork lift truck (curtain-siders would have been the natural choice had they been invented in the early 60's).

The trucks and trailers were painted in deep crimson and proudly carried the Vauxhall name and logo. By the end of 1964 28 loads per day were moved. Eventually the planned capacity of 50 tractor units and 112 trailers making 48 return loads per day would be met. In actual operation the trucks invariably reached full capacity before the weight limit was exceeded.

How Cars Were Made between 1963 & 1966

Production Methods

In industry there are three basic scales of production: one-off, batch production and mass production.

One-off (or craft) production is where only one or a very few examples of an item are required. Here the parts are produced by skilled craftsmen on standard machine tools which are set up to make one or just a few of the items, and each part is moved around the workshop to the various machines, e.g. lathes, drilling machines, grinding machines etc. This method is usually found in tool rooms or small jobbing workshops.

Batch production is used when larger quantities of parts are required from time to time. Here the parts are manufactured using standard machine tools but often using special jigs or fixtures designed for the particular part. This method eliminates some of the skills needed to achieve the necessary component accuracy and consistency. Vauxhall Motors used this method for the production of non-current spare parts. Mass production techniques are used when large quantities of parts are produced continually to feed a major unit assembly line. The machine tools are laid out in the sequence dictated by a production planner, each machine being set up to carry out one operation before the part is transferred, often automatically, to the next operation. The machines are laid out and set up on a permanent basis. The accuracy of the parts produced is controlled by the machines and the dimensions are checked for size using purpose-designed Go/Not Go gauges. Each machine line is usually laid out in a manner where the parts end up nearest to the point on an assembly line where they are fitted to the major assembly.

To assemble the Viva engine the cylinder block was bolted on to an adjustable pedestal which was carried on the floor level continuous conveyor. The engine's components were fitted to the cylinder block. In figure 64 the engine has reached the stage where the gear box will be fitted. It would then have been removed from the pedestal before final testing.

Whilst Ford is popularly cited as the originator of mass production, they were a comparatively late developer of this technique. The idea was first developed in Venice several hundred years earlier, where ships were mass-produced using pre-manufactured parts and assembly lines. The Venice Arsenal apparently produced nearly one ship every day, in what was effectively the world's first factory which, at its height, employed 16,000 people. This system developed significantly during the second half of the 19th century, particularly in the USA, where large numbers of goods such as sewing machines, firearms, bicycles etc., demanded by a rapidly expanding population were being produced by a relatively small and often unskilled labour force. Mass production was developed by firms such as Singer for sewing machines and Smith and Wesson for firearms. One key development was the interchangeablity of parts which was essential for unskilled assembly work. It also meant that replacement parts could be fitted retrospectively without the need for any adjustment to size in order to fit and work properly. However, it was Henry Ford who took the system to its peak for his Model T car in the first quarter of the 20th century.

The inflexibility, inherent with special purpose machines used in mass production, means that long production runs are essential to recoup the large capital investment.

Figure 64: Batch production of non-current spares on an Alfred Herbert multi-pillar drill.

Figure 65: Viva engine assembly conveyor at Ellesmere Port.

Figure 66: An operator tightens up wheel nuts with a multi spindle compressed air powered nut runner.

Mechanical Components

At Ellesmere Port, Vauxhall used standard machine tools with dedicated jigs and fixtures wherever possible, and with specially developed automatic handling of components between these machines. This system minimised inter-operational banks of components (work in progress) and smoothed production flow, at the same time reducing the labour cost of components.

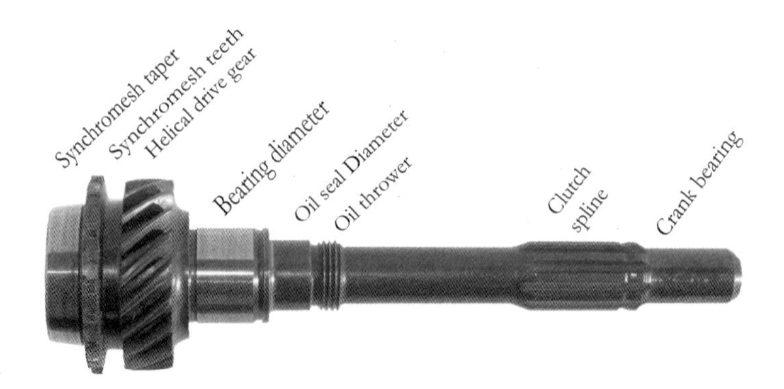

Figure 67: HA Viva Main Drive Pinion part number 6376016.

To illustrate the principles of mass production in practice, the list below is the sequence of operations made to produce the gearbox main drive pinions:

1. In the heat treatment department the forging is "normalised" to remove forging stresses and shot blasted to remove scale.

2. The forging is faced and centre drilled at both ends, one machine used.

3. The stem is turned in a copy lathe up to the back face of the gear diameter. - 1 machine

4. The gear outside diameter, clutch teeth diameter and synchromesh taper are turned and internal bearing is bored out on a six spindle automatic lathe. - 1 machine

5. The oil thrower is screw-cut in a "Cridan" lathe. -1 machine

6. Three oil feed holes behind the clutch teeth are drilled. - 1 machine

7. The synchromesh clutch teeth are gear shaped.- 3 machines

8. The helical gear teeth are rough gear shaped. - 6 machines

9. The helical gear teeth are finish gear shaped. - 6 machines

10. The spline for the main clutch is milled. - 1 machine

11. The synchromesh clutch teeth are tooth chamfered. - 1 machine

12. Helical gear teeth are de-burred and chamfered on the acute angles at each end. -1 machine

13. The helical drive gear is shaved. - 1 machine

14. Wash to remove all cutting oils and swarf fragments.

15. Case hardened all over in the heat treatment department.

16. Synchromesh taper is ground to size.- 1 machine

17. Bearing and crank shaft bearing diameters are ground to size, 1 machine

18. Internal bearing diameter for main shaft bearing is internally ground to size. - 1 machine

19. Wash to remove all grinding fluids, abrasive particles and dirt.

20. Final inspection and pass to assembly area.

Each machine tool was positioned in a line as dictated by the operation sequence and components were automatically transferred between each operation on free flow conveyors. This system allowed for a small bank of components between each operation so that, if one machine was stopped for tool changing etc., the whole line did not stop. In the example above most of the machine tools were automatically loaded.

Quality Control

All component dimensions were checked as required using either simple gauges of the Go/Not Go type or special purpose fixtures such as the one below. The use of these gauges eliminated the skilled use of standard measuring equipment such as micrometers or verniers.

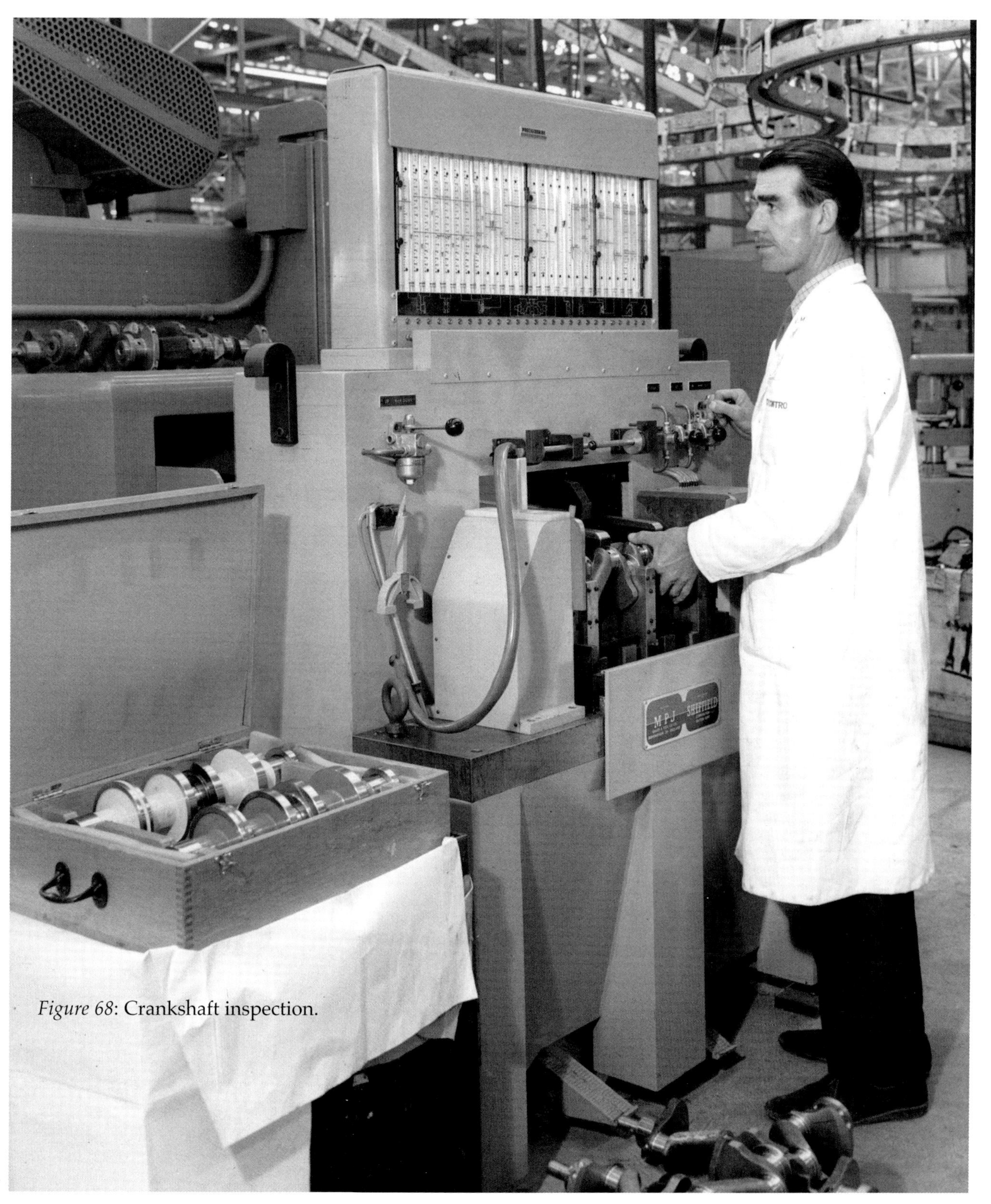

Figure 68: Crankshaft inspection.

The illustration on the previous page shows an inspector checking 28 critical dimensions on a crankshaft using a multi-point air gauge. This was capable of measuring crankshaft bearing diameters to 0.001inch (one thousandth inch) tolerances. The picture was taken early in 1963 during the proving stages for the Viva engine and before the production storage racks were available. It was certainly not standard practice to leave finished crankshafts lying on the floor.

Transfer machines

For complex components, such as cylinder blocks, cylinder heads or gearbox cases, transfer machines produced many thousands of one particular component. The component casting or forging was pre-machined with surfaces and holes for primary location. They were then loaded into the transfer machine. The part then progressed one station at a time past a succession of machining heads necessary to carry out the various milling, drilling, reaming and boring operations before being unloaded as a finished component.

Figure 69: This Ingersoll transfer machine was used to completely machine every Viva and Chevette cylinder block between 1963 and 1985, over two million components.

Car Assembly

The car assembly method of production varied considerably from the methods used in the production of the mechanical components (i.e. engines, gearboxes and axles.).

The whole of the car assembly operation was carried out in ED Block, where rolled or cut sheet steel was received at one end and finished cars were driven out at the far end of the building. The mechanical components were delivered by an overhead conveyor from EA Block in the correct sequence for the marriage conveyor (see figure 74).

Press Shop [62]

The major difference between the press shop and the rest of the factory was that it operated on a batch production basis because the hourly output from the presses was at a far greater rate than that needed by the assembly track (conveyor). The press tools for each component were set up in a press of appropriate size and a batch of several thousand pieces was produced and stored until required for sub-assembly. Each press used a pair of steel dies, an upper and a lower, between which sheet steel was pressed into shape. Several operations would be needed for the more complex components.

Starting in 1964, from small beginnings in EA Block, the Ellesmere Port press shop story began. In late 1964 and on into 1965 small presses were installed in EC Block, some small presses called "inclinables", together with two larger 80 ton presses, which were transferred from Luton to start the manufacture of small parts for axle and gearbox assemblies. The largest component produced at that time was the rear axle housing cover, which was about nine inches diameter.

The initial phase allowed for training operators, setters and their supervisors in press operations. This was done prior to transferring operations into the all new ED Block press shop. The press shop was designed to manufacture, from "as received" sheet and rolled steel, all the body panels for the new HB Viva which commenced production in 1966.

The design of the press shop had the presses mounted at ground floor level with a large basement area underneath. Here scrap steel was funnelled down chutes to conveyors feeding the sheet metal baler, were it was squeezed into compacted blocks of scrap ready for recycling at the steel mills.

Figure 70: Rolls of sheet steel being checked after receipt from John Summers and Sons at nearby Shotton.

Steel was received direct from the steel mills as either rolls, with up to a kilometre length of steel on each roll, or as cut blanks. These were all stored inside the press shop basement, from where they were lifted by crane to the pre-operations. These involved flex rolling the steel from rolls to remove manufacturing stresses and then cut into the required lengths. This flex rolling process was necessary to prepare the sheet and thus helped prevent splitting during press operations. Particularly deep pressings were greased to aid deep drawing and prevent splitting. Deep pressings were always prone to the metal splitting rather than flowing to the required shape.

The main press shop had fourteen major press lines, but the number of presses in each line varied. On the smaller press lines not all the presses were necessary to produce some parts. On these types of lines, the parts were hand loaded into the dies. The parts were then unloaded onto a belt. At the end of the line the parts were put into a stillage or a rack for storage. On the major lines the largest parts were produced, such as body side inner and outer panels, under-bodies and roof panels.

[62] *Press Shop* section written by Hilbert Talbot

Figure 71: Press shop showing a line of nine presses.

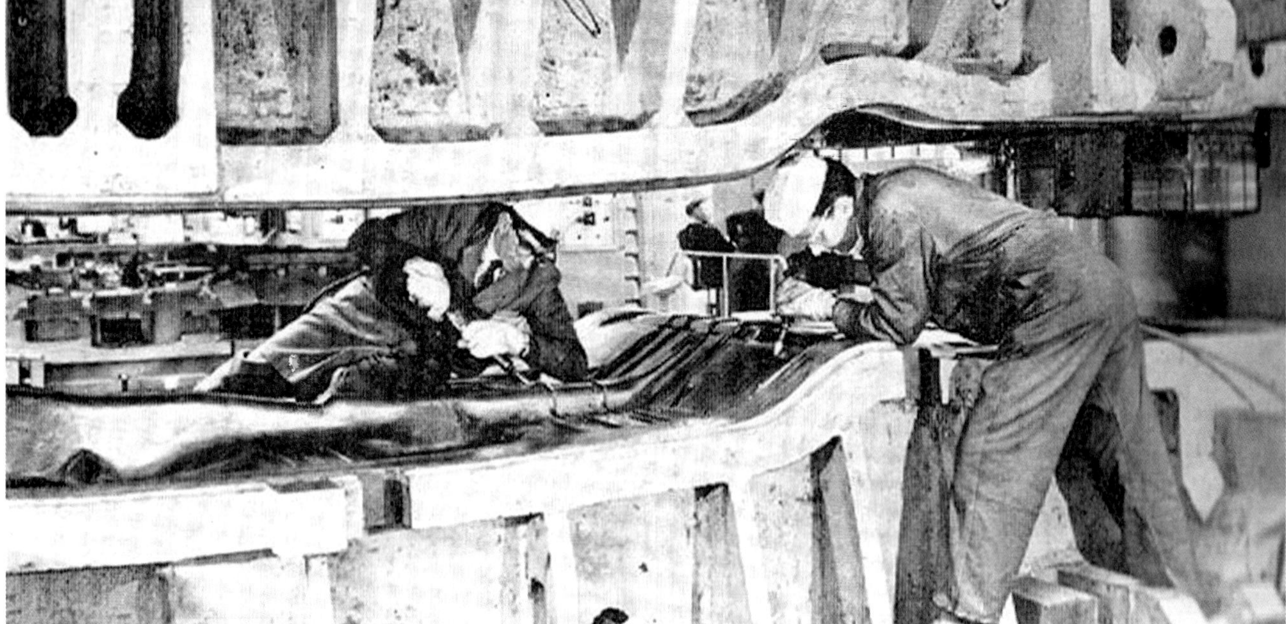

Figure 72: A pair of dies for pressing floor panels in one of the 1,000 ton presses. Two die maintenance fitters are making small repairs to the lower die to improve the quality of the pressings. Steel blocks were used to prevent the dies accidentally closing and causing catastrophic injuries to the maintenance fitters.

These 1,000 ton presses were also hand loaded but automatically unloaded using an "iron hand" onto a belt for passage to the next press. The final operation saw the finished parts unloaded onto racks ready for the body shop.

Press shops can be very dangerous places and safety was always a major consideration. This was achieved on the presses themselves by using safety interlocks. These prevented the presses operating when they were open for loading and unloading. A safety bar, which pushed the operator away, had to be fully extended or the operator had to press two widely separated palm buttons simultaneously to start the press. Either method ensured the operator's safety.

ED Block Layout

From the body shop onwards all the major assembly conveyor tracks ran across the width of the building, before turning 180 degrees and returning again in a giant firecracker layout. This arrangement was adopted so that the conveyor length could, relatively easily, be increased to accommodate extra operations. With this layout a section of the western wall of the building could be moved out and a loop of the conveyor lengthened into the new, enlarged floor space.

Body Shop

From the press shop the body panels passed into the body shop where the car body was made up of six major units – the floor-pan, two sides, front and rear ends and the roof. There were a number of sub-assembly areas where the various sections were built on their own conveyors. All these subsidiary conveyors producing the underbody sections, sides etc. were synchronized with the main body build conveyor, which they fed.

The main body weld conveyor on which the body shells took shape carried 32 special jig trucks on a large oval circuit. The main body units were clamped together on these trucks while the spot welding was done, the roof being the last major unit to go on. This was known as the "gate line". Spot welding alone does not produce a watertight body; most joints were sealed using special compounds.

From the gate line conveyor the body shells were transferred to another conveyor where doors, bonnets and boot lids were fitted. Various forms of welding were used to build up a car body: spot welding, seam welding etc. A single car body could contain up to 5,000 individual spot welds. At the end of the assembly process the bodies were then inspected and surface blemishes removed prior to going to a small holding bank ready for painting.

Paint Shop

The main paint shop was at first floor level above the main assembly areas. Here the car bodies were degreased before starting the long and thorough paint process which meant more than just giving them a coat of bright shiny colour. First there was a seven stage rust proofing process using alkali and hot water washes at 140°F, a phosphate spray and a chromic acid rinse before drying at 325°F. The next stage was almost complete immersion into a 5,000 gallon bath of black primer paint. Dipping the body forced the paint into every nook and cranny in the body shell. Then the body received a two coat spray of primer paint. This was applied by automatic spray guns while hand guns were used for the inaccessible areas such as wheel arches and door shuts. These coats of paint were dried off at about 325°F.

Figure 73: Paint shop wet sanding area in the late 1960s with HB Vivas.

After wet sanding the paint, the bodies were again dried before receiving three coats of the finish colour and a hard wearing acrylic lacquer. This was dried and then the car passed through a thermal reflow oven at 320°F which brought out the lustre on the paint.

Wax fillers were sprayed into door sills and the underbody was sprayed with a sealer/sound deadening compound.

Hard Trim

After painting the body shells were passed into another small storage bank prior to being finished in the two trim shops. These banks allowed for non-coincidental shift start and finish times. They also absorbed any small hold-ups in departments before or after the banks.

In the hard trim shop the body was moved on a lightweight wheel trolley where, at over eighty stations, various trim items such as glass, instruments, electrical wiring and lights were fitted. At this stage the marriage of the body and the mechanical components took place. The purpose was to produce a functional car which could be inspected and tested before soft trim.

Figure 74: The final stage of the marriage conveyor.

The final assembly was done with the body well above floor level. For a short distance an auxiliary conveyor ran beneath the overhead conveyor on which the bodies were suspended. This auxiliary conveyor carried special trucks on which the front and rear axles and engine-gearbox assemblies were mated together. These trucks were provided with jacking mechanisms so that the linked mechanical units could be connected to the car body. This is shown in figure 72. As the cars descended towards floor level they met another conveyor carrying wheels and tyres already correctly inflated. At the end of this "marriage conveyor" the car was put on a dual track conveyor at floor level. At this stage oil, water and fuel were added ready to run the car on rollers.

Here the cars were "driven" in all gears, and the brakes, lights and instruments were all checked by inspectors. After acceptance by the inspector, the car was tested in a high pressure water-jet tunnel to ensure that it was leakproof. At this point we had a fully functional and waterproof car, although not a very comfortable one! We just needed seats etc. – soft trim.

Figure 75: An inspector is checking in the mirrors to ensure that the front and rear lights are all working correctly

Soft Trim

The final assembly area was where soft trim items such as headlining, carpets, door trims and seats were fitted before the final inspection and setting the headlamp beams. The cars were then passed to the sales department for distribution to UK dealers or shipping agents for export.

Cars Produced at Ellesmere Port since 1964

The first car built in Ellesmere Port was the HA Viva. It was produced from June 1964 in EC Block on a low volume training pilot line in readiness for full-scale production, when ED Block would become operational in 1966.

There was one basic body shell; a two door saloon with only one engine specified, an 1157cc petrol engine. Although there were two trim levels, the standard and deluxe, the only added features were a heater for passenger comfort and a windscreen wash.

When ED Block came on line in 1966, it was with the HB model Viva. This was a very pretty car and was eventually available in either two or four doors versions and an estate car. A higher performance engine was available with 60 BHP instead of the standard 47 BHP.

In 1970 the HB was replaced by the HC Viva. This had a considerably larger body and a 1256cc engine. The HC was produced with the same three body styles as the HB. In addition the Magnum versions had either 1800 or 2300cc OHC engines.

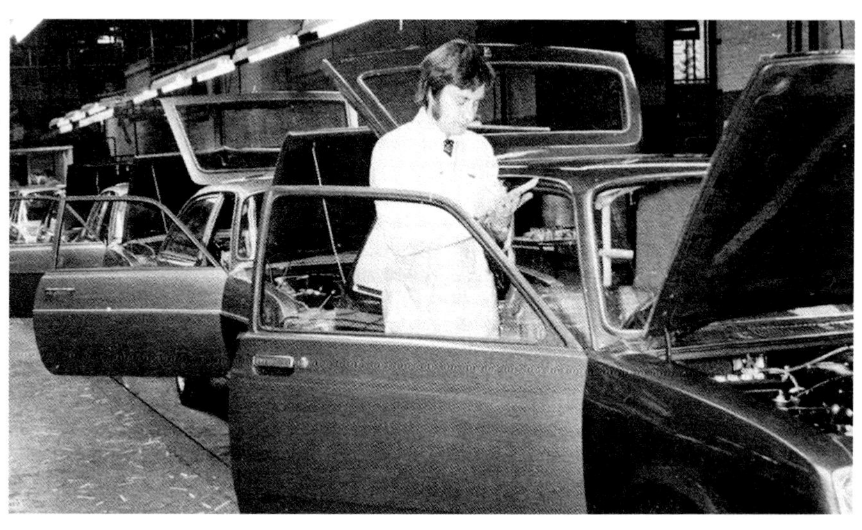

Figure 76: Cars receive their final inspection before leaving the assembly building.

You will enjoy these sporty qualities in the all-new Viva 90!

	REGULAR VIVA	VIVA 90
Max. net b.h.p.	47 at 5,200 r.p.m.	**60 at 5,600 r.p.m.**
Max net torque	62 at 2,800 r.p.m.	**64 at 3,600 r.p.m.**
Brakes	Drum brakes, 8 x 1¼ in.	**Power brakes (front discs).**
Tyres	Tubeless 5.50 x 12, 4 ply rating.	**Tubeless 'ultra-low profile' 6.20 x 12, 4 ply rating.**

And you get these 'extra' extras with the all-new Viva 90 SL

Performance, luxury, beauty of line, wonderful Viva gearbox, all-new coil suspension, space-curve shape, wide-track design—you get all these in the Viva 90 de Luxe and Viva 90 SL. But the Viva 90 SL gives you *extra* extras too. Superb seating in resilient Ambla. Rear seats with a luxury 'individual' look. Cut-pile carpeting. Extra sound insulation. A beautiful car with distinctive grille and wheel trim. Viva 90 SL! Talk to your Vauxhall dealer now and test-drive these exciting Viva 90s for yourself.

All-new Viva 90 by Vauxhall

Viva 90 de Luxe £663
Viva 90 SL £708
Prices include P.T.
Delivery charges extra

Figure 77: An early advert for the HB Viva focussing on the high quality upholstery.

HIGHLIGHTS OF TEN YEARS

NOVEMBER 1962
Initial transfer of some machining operations from Dunstable to Ellesmere Port.

First locally engaged employees joined the company at Ellesmere Port.

JANUARY 1963
EA Block essentially complete – as were the boiler house, switch house and trade effluent plant.

JUNE 1963
Second phase of development started when work commenced on building EC Block.

AUGUST 1963
HA Viva introduced – with complete machining and assembly of engines, gearbox and rear axles at Ellesmere Port.

MAY 1964
Third development phase started – ED Block which would include press, shop, paint, trim and final assembly facilities. On completion would represent a total investment of £30 million.

JUNE 1964
Temporary facilities for the assembly of HA Vivas established in EC Block.

NOVEMBER 1964
Vivas were being assembled at a rate of eight per hour. Total assembled at the plant had reached 5,500.

FEBRUARY 1965
Announced that 1,000 new jobs would become available over next few months as ED Block was brought progressively into use.

MARCH 1965
First locally-made Vivas to be despatched rail. Consigned from Hooton Station to Bathgate (Scotland).

JULY 1965
First export consignment from Ellesmere Port docks – 900 cars shipped to Canada aboard MV "Madame Butterfly".

JANUARY 1966
Work started in new Viva assembly building (ED Block) and press shop also started to come into operation.

SEPTEMBER 1966
HB Viva announced – manufactured at the Ellesmere Port plant.

NOVEMBER 1966
Double-shift working started in paint shop, body-shop, trim and final assembly.

JANUARY 1967
Automatic transmission became available on some models in the Viva range.

JUNE 1967
Viva Estate cars were added to the range.

MARCH 1968
The high performance 2-litre Viva GT became available.

OCTOBER 1970
The HC range of Viva saloons and estate cars replaced the HB series.

MAY 1971
Celebrations to mark the production of one million Vivas were held at the Luton and Ellesmere Port plants. At Luton, Sir John Eden drove the millionth Viva off the line, whilst at Ellesmere Port Ken Dodd officiated when the first of the second million was produced.

JULY 1971
Dealer Team Vauxhall established for racing and rallying Vivas and Firenzas. Four championships won – followed in 1972 by nine championship successes.

MARCH 1972
Increased engine size for Vivas and Firenzas.

NOVEMBER 1972
Tenth Anniversary Celebrations.

The three generations of Vivas which have helped the Ellesmere Port plant towards world-wide fame–(left to right) the HA Viva de luxe (1963-66), the HB Viva SL (1966-70) and the current HC Viva SL. Together they have made the Viva the most successful range of cars in the company's history.

Figure 78, upper right: On the tenth anniversary of the plant opening the three generations of Vivas are lined up during the birthday celebrations. The HC model has the fashionable vinyl roof!

Figure 79, left: 1972 saw the celebration of ten years of Vauxhall at "The Port" and notable events were recorded in a special editon of the *Vauxhall Mirror* published in November '72.

Figure 80, right: HC Viva with a 2.3 litre engine. This advert emphasises the sporting and family attributes.

1978 saw the beginning of the end for Vauxhall-designed cars. The assembly of the hatchback Chevette coincided with General Motors Europeanisation policy for the design and manufacture of cars. Basically it was an Opel Kadette with a restyled nose. This was the last car to use the Ellesmere Port manufactured 1256cc engine, which was by now twenty years old and was technically behind the times. All subsequent small and mid-sized cars were front wheel drive and used more modern transverse overhead cam power and transmission units, which were built at other GM plants. This change sounded the death knell for mechanical component manufacture at Ellesmere Port.

Figure 81: A 1982 Astra Van publicity shot.

1979 saw the introduction of the new GM "World Car", the Astra hatchback. It was produced in both two door and four door versions. However, the marketing department liked to count the tailgate as a door, so called them three and five door cars. Vauxhall, now utilizing GM's worldwide manufacturing facilities, were able to offer all body variants with a choice of 1.4 litre, 1.6 litre petrol engines and a 1.6 litre diesel engine, all with 4 or 5 speed gearboxes. There were also estate car versions and in 1982 a Bedford van variant was offered. This took the market by storm. It was the first time customers were offered a car-derived van with a diesel engine. This engine was expected to take about 30% of the van production but very soon it reached 70%. In the first year of production the Bedford van achieved market leadership in its class, surpassing the traditional leader, the Ford Escort van.

After 29 years of Astras, the plant is now producing the sixth generation of cars and vans. The permutations of three body styles, seven engines, two gearbox alternatives and six trim levels, not to mention "What colour would you like?", are mind-boggling. Compare this with the HA Viva which, at the launch date, offered only one engine, one body style and two trim levels. But the marketing gurus keep saying that in order to retain a strong presence in the market, it is necessary to provide the variants that the customer wants.

Working conditions in the plant

Vauxhall's aim was to provide a clean, healthy and safe environment for their employees and in general this was always to a high standard. Space heaters installed in the roof trusses provided heating and all gangways and floor areas were regularly swept to clear any swarf and litter. However, over the years many of the machine tools gradually developed oil leaks. This leakage was absorbed by spreading fine cast iron swarf on the floor. Cast iron, because of its high carbon content, is naturally dirty and one's shoes soon became oily and dirty. Woe betide you if you went home in your work shoes!

Steel swarf was often produced in short lengths in a figure "9" shape. This was also sharp and cut the rubber soles of shoes so they tended to split across the ball of the foot. Production operators were offered wooden clogs, but these were neither comfortable or practical if you had a "walkabout" job.

During hot weather the smoke produced by metal cutting operations was not always cleared by the roof ventilators and hung as a smog over the machining areas. The noise level, whilst within legal limits, meant that speech was not easy in some parts of the factory and people unconsciously always faced the person they were speaking to and partially lip read. Ear plugs or ear defenders were provided in the press shop.

In the body shop, solder puddling (a tin and lead alloy) was applied to conceal some of the body panel joints. Workers on these activities were given a pint of milk every day to absorb any lead and regularly checked for lead levels in their blood. If the level was too high they were moved away from the lead fumes to another job.

Trade Union Activity Up To 1980

Wages for hourly paid employees at Ellesmere Port were initially to be sixpence per hour less than for the equivalent grades at the southern plants. This was said to recognise that the new starters had little or no experience in the motor industry and would therefore be less efficient. This differential was scheduled to be fourpence per hour in year two and twopence per hour in year three. This policy was also to appease local employers who were afraid that the motor industry's traditional high wages would upset their own wage agreements. However, trade unions soon started to press for the difference to be eliminated. Arguments by the unions that it did not take three years to become an efficient labourer or toilet cleaner were not easily refuted by Vauxhall management. Due to trade union pressure the north/south differentials were abandoned after about six months (Ford also lost their differential of 1s. 3d. in 1963 due to trade union pressure). This was an early example of the effectiveness of trade unions. The defeat cost Vauxhall about £240,000 over the first three years and gained the unions a lot of credibility as effective defenders of working conditions. Their power and influence increased to the extent that the Management Advisory Committee, which had been introduced fairly early at Ellesmere Port, soon lost influence because the unions were seen to be more effective. So, after being boycotted by shop stewards for a few meetings, the MAC was disbanded.

The trade union movement at that time was suspected of being politically motivated to cause trouble in British industry and the motor industry in particular. Vauxhall was seen as a soft target. Industrial strife continued with many disputes and much lost production not only at Ellesmere Port, but also at the southern plants because Ellesmere Port failed to supply the components essential to keeping the tracks running. By 1979 GM realised that they either had to regain control of the plant or close it. That year the hourly paid employees voted in favour of a vague mandate to allow union officials to "Take whatever action they felt necessary to get a pay rise". The union used this mandate to call a strike. After twelve weeks the unions held a vote with an ambiguous result to continue the strike. However, the company discovered that many workers were no longer supporting the union tactics and wanted to return to work.

The company took the unusual step of inviting workers to ignore the vote and return at their own initiative. The first day about 600 of the 10,000 employees were invited back to work. On their return, every employee was read a document by his foreman entitled "We Will Manage". This outlined the new rational working agreements. It swept aside all previous "local" agreements and virtually imposed a new plant-wide agreement on working conditions and common sense standards of performance, re-establishing the long-term Vauxhall principle of "paying a fair day's wage for a fair day's work". Employees were given the stark choice of either accepting these conditions or going back home! After twelve weeks with no pay (because of a dispute in which many felt they had been tricked by the union into striking) most agreed to the new terms and started work. On the second day a larger group (about 1,200) were invited back to work and were given the same choice, accept or go back home. This group included a few shop stewards, who rejected the new agreement and tried to take their members out with them. Generally they received no support and were forced to accept a fait-accompli. After a few more days all employees had returned to work under the terms of the new plant-wide agreement and the trade unions had lost their stranglehold on the plant.

After this event production gradually returned to normal: output and efficiency increased far above the levels achieved previously[63] when there had been frequent disruption due to unofficial stoppages and consequent loss of production. Charlie Golden, Chairman and M.D. 1993-1996 said, "I was a finance analyst at the New York office in 1979 and few people in the UK can have realised then just how close Vauxhall came to being shut down during the dispute. The recovery afterwards really was a last chance for the company"[64] This period from about 1965 to 1980 would make a suitable subject for study by others more competent than the present author.

By the mid-1980s Vauxhall small and medium sized cars had changed from being rear wheel drive to transverse engined front wheel drive cars following the trend set by the BMC Mini of 1959. This change necessitated a brand new engine and transmission design to be used across the whole range of GM's smaller cars in Europe. However, it was not to be built at Ellesmere Port, as by that time GM had a policy of worldwide sourcing for major components. The original Viva engines, gearboxes and axles were no longer required. This meant that most of the machine tools and assembly lines in EA and EC Blocks were redundant and these were sold or scrapped.

[63] *Ellesmere Port Pioneer*, February 1980

[64] Informal discussion with Charlie Golden, October 1994

By 1990 Ellesmere Port's productivity record and labour relations reputation had improved sufficiently for the plant to win a bid, in partnership with Fiat of Italy, to build a new V6 engine. The V6 engine shop occupied the floor area of EA Block, previously used for both engine and passenger gear box machining. The new engine would be supplied for use by GM and Fiat worldwide. GM had, at that time, taken a share in the ownership of Fiat SpA of Italy and the V6 engine was an early stage in the long-term plan for product integration.

Biographies and Reminiscences

Hazel Alston (secretary)

Hazel was secretary to Eric Fountain in January 1963 and her memories of when the main EA administration and production block was being built centred around gum boots, conveyors overhead and holding body and soul together during the raw winter. "It was so cold we had to wear gum boots and overcoats in the office". She remembers how shrewd she was forced to be as contractors would call in for a chat and try reading tenders upside down in her in-tray. "There was so much contract work going out to tender I even had to shred my carbon papers". Writing in 1987 she was "proud to be the longest serving locally recruited female employee."

Tony Burnip (production foreman)

Tony had been a Vauxhall apprentice at Luton and was appointed foreman in the Luton engine shop in 1961. He transferred to Ellesmere Port in January 1963 to work on the finish grinding section in the passenger car gear shop. Several promotions over the years saw him as works manager before he was appointed plant manager in December 1979. In April 1986 he was appointed as director of manufacturing, GM Espana in Zaragoza in Spain. In 1990 he became Vauxhall's supply director and a member of the Vauxhall board of directors.

Eric Fountain (planning engineer, factory layout)

Eric joined the company as an apprentice in 1943 and was the chief engineer in Vauxhall's factory layout department, when in 1959, with the project manager Tom Williams and Sir Reginald Pearson they spotted Hooton Park Airfield on their way to a meeting with the Ellesmere Port town clerk to discuss possible sites. Eric then worked full-time planning the Ellesmere Port plant until he transferred to the plant in the autumn of 1962 in a similar capacity. In 1968 he moved back to Luton to be appointed to a number of senior production management positions in car assembly operations. Two years later he returned to Ellesmere Port as director and plant manager. He was the first plant manager to be a main board director of the company. A further change in career saw him as director for public affairs. He retired from this position in 1989. The contribution Eric made to the plant and the community at large has been recognised by the award of the O.B.E in 1987 and the naming a road to the new Ellesmere Port supplier park in January 1989.

Writing in 1982[65] Eric recalled some of his early days: "The pioneering spirit was incredible. Nothing was too much trouble for anyone. Conditions were quite awful – we went through a very bad winter in the early stages of construction but morale was fantastic – that is why Ellesmere Port in the early 1960s was such a success. Industrial problems were to come later – and Ellesmere Port went through a ten year "bad patch". In those early days disagreements between management and the workforce were less fundamental. The first gripe was because we couldn't make enough chips in the canteen. We underestimated the liking for chips in the area – our canteen staff were only geared up to make four tons of chips a week. Once we had altered things to make seven tons a week, things rapidly got back to normal."

Vic Honey (production operator, non-current spares)

Vic was one of the first group of production operators who joined the company on 12th November 1962. Previously a machine tool setter at Girling Ltd. in Bromborough, he became the first local recruit to be re-graded as a machine setter in the commercial gear shop early in 1963. Soon afterwards his talent for dealing with people made him the first of the local recruits to be appointed as a foreman. After a few years he was promoted to general foreman in the commercial gear shop.

[65] *Chester Chronicle* supplement to celebrate 20 years of Vauxhall at Elesmere Port, January 1982

In 1982 he volunteered for a completely different assignment: training dealership staff in the operation of a voice response computer terminal. This was the first time dealers would be able to have direct access to Vauxhall's computer systems. It would enable them to order vehicles and spare parts on-line and to carry out various associated activities, such as checking the new vehicle status in the production sequence, or locating vehicles already in stock at other dealerships.

This was always intended to be a short-term assignment, but very soon afterwards another opportunity arose to lead a team of nine foremen from Ellesmere Port, who were recruited temporarily to help dealers improve their van sales. After about ten months this task force was transferred permanently to Bedford marketing division. Vic then was appointed as a sales development manager for Bedford bus and coaches. Here he liaised with dealers and customers to facilitate and improve sales. It was from this position that he took early retirement in 1987. Vic's story illustrates how someone could develop their skills and progress in various directions, in his case, firstly his practical skills with machines and then his management and people related skills.

Roy Hooley (machine setter)

Roy also joined the company on that first day. He claimed the honour of being the first person to actually produce a component at the plant, a gearbox mainshaft which would be supplied as a non-current spare to the parts department at Dunstable.

Roy remained a machine setter on gear shaving machines for his whole time at Vauxhall and retired after about 25 years service. In retirement he became a gardener at Chester Cathedral. In the picture below, taken on the tenth anniversary of the plant opening, Roy is showing the plant manager one of the components he produced in 1962.

Figure 82: Roy Hooley and Rob Walker celebrate ten years of production at Ellesmere Port.

Mike Smith (production foreman, heat treatment)

"Originally from Chesterfield in Derbyshire, I joined Vauxhall at Luton in September 1957 as a five year engineering apprentice after which I was graded as a production technical assistant. In March 1963 I transferred to Ellesmere Port as a production foreman. The transfer package was 20% of salary paid as lump sum for married staff buying a house near Ellesmere Port. This was to cover the cost of carpets, soft furnishings etc. Unmarried staff received 10%. These payments were in addition to the company paying all removal, estate agents and legal fees. All staff received £13 per week for thirteen weeks to cover temporary accommodation etc.

Initially I was assigned to the heat treatment area as a foreman. This area was claimed to be the largest such facility in the UK and was a continuous process needing management cover 24 hours a day. The foreman covering the "dog hours" between production shifts was also responsible for any other production activity. One evening I heard fire sirens and saw emergency vehicles pulling up outside the plant. Being aware of the devastating effect if the nearby large propane tanks exploded, I quickly found two men in suits and bowler hats with clip boards who seemed to be involved. They informed me that they were from the Home Office and this was an exercise carried out twice a year to verify that the Fire Authority had the procedures in place to deal with such an emergency.

One thing we quickly learned about Merseysiders was their sense of humour and their talent for giving their foreman a nickname based on his favourite expressions, to list just a few:

The Sherriff	"Where's the holdup lads?"
The Drug Addict	"When you've done doze I've got more fer ye." (Morphia)
The Balloon Man	"Don't let me down lads."
The Film Producer	"Let me put you in the picture."
The Gardener	"I want you to plant yourself there."
Rommel	always phoning from Car Assembly "I must have more (petrol) tanks."
The Margarine Man	"Spread out lads."

And names for fellow workers:

The Crab	"I won't be in today, one of my nippers is ill."
The Destroyer	He's always after a sub.

I later became an inspection foreman in the commercial gear shop. One of my pals from our apprentice days was **Neil McKechnie**. Originally from Wallasey he came to Luton, like myself, for his apprenticeship. He transferred to Ellesmere Port at the end of 1962 as a production foreman in the non-current spares area. Neil was already well known on the Wirral as a member of Wallasey Swimming Club. He swam for Great Britain in the 1956 Melbourne Olympic Games and two years later gained a bronze medal in the 4 x 200 medley relay at the Commonwealth Games in Cardiff. Neil was predominantly a 400 metres freestyle swimmer but in 1958 he held every English freestyle record from 100 yards to one mile.[66] He had a huge frame and regularly polished off two dinners in the canteen. Neil left Vauxhall around 1970 but always remained loyal to the company and, later, often brought groups of his employees to Ellesmere Port on factory visits. Sadly Neil died following a heart attack in June 2006.

Figure 83: This is typical of the houses, built by Wimpey Homes in the 1960s around Ellesmere Port, which were bought by many employees new to the area.

[66] http://www.wirralglobe.co.uk/news/792629.0/ *op. cit.*

Eddie Thomas (production operator, passenger gear shop)

"I first joined Vauxhall late in 1962, This was after serving an apprenticeship with D. Napier and Sons on Merseyside, then a further five years as a skilled machinist producing parts mainly for their Deltic marine and rail traction engines. Napier was a company renowned for its very high quality and precision engineering. By that time I was a setter-operator on a Maag gear machine grinding the flanks of very large gear teeth. This was recognised as the most skilled and highest hourly paid job in Napiers. With this background I, and a friend Bert Pearsall, filled out application forms for interview at Ellesmere Port. To the question – "Present Position" we both entered "Gear Grinder", naively believing that the term was universal and denoted a high skill level. I believe it was Arthur Kaye, engine shop area manager, who interviewed us. He seemed to be bored stiff after conducting similar interviews for several days, but accepted my application as a 'gear grinder' and we were given a starting date. Subsequently, Bert and I reported for work at the new plant. At that time there were no walls to the factory building, only tarpaulins stretching around the outside pillars to keep out the wind. The whole place gave the impression of being in a shambles with very few machines installed and certainly no gear grinders for me to operate! We were amazed when we were given the widest yard brooms we had ever seen and told to sweep the aisles! Neither of us spoke for a while as we worked alongside one another, brushing up straw, sawdust and other rubbish, then Bert turned and said breathlessly, "Somebody's made a mistake here." I almost collapsed laughing and replied, "But we're earning more than we ever did for so much less responsibility." We brushed the floor for three days and laughed the whole time. When the centreless grinders were installed I was made a machine operator but soon realised that I could not live with the situation, regardless of the pay. I was told that I was one of many who had misunderstood that it would take some time before we could be reassigned to jobs which more closely matched our previous experience. I waited six months before quitting to go to the new Ford Transmission Plant at Halewood in their Gear Development department. Bert became an inspector in the passenger gear shop and stayed at Vauxhall for many years. It is a very sobering thought that, had I stayed, I would probably have ended up sharpening cutters in the tool room!

In 1965 I applied again to Vauxhall, this time for a technical vacancy and was recruited as a tool trials and test engineer responsible for cutting tool development and later became a planning engineer in the gear shop.

After 15 years I left Vauxhall and emigrated, initially to join Western Gear Corporation in Los Angeles, and later as a consultant with Ingersoll Engineers, of Rockford, Illinois. In about 1986 I was assigned to a project working with a consortium of engineers at the GM Advanced Engineering and Research Department in Detroit, Michigan. Each month I would attend a meeting at the main GM plant in Michigan. After one of these meetings a group of us were sitting in the conference room having lunch. Apart from myself all the others were GM employees. The chairman pulled out a notepad and said, "Let's have some fun. Each of you tell me how many years service you have with GM and I think we'll be surprised with the amount of expertise around this table". He pointed at the guy on his right who said, "18 years" and so on round the table, ignoring me completely. I said nothing until he finished his calculations and announced, "137 years - isn't that amazing?" At which I said, "Plus 15". Everybody turned to look at me and one guy asked at which GM plant I had worked. At my reply there was a sharp intake of breath all around the table. One of them explained that they were all aware that, some years ago, Ellesmere Port was only two weeks away from being closed down completely. "The workforce was no good", said the chairman, "Those people would strike for no reason whatsoever." I replied, "You guys need to know something about the Merseyside people. Directly across the River Mersey is the Ford, Halewood Plant. This is the only place where Ford have beaten GM in sales in any country. It is referred to as Ford's flagship operation. These are the SAME PEOPLE. Many had worked at both Vauxhall and Ford. So what conclusion can you reach? Here's a clue – Ellesmere Port had suffered from a change of plant manager almost every 18 months since its inception. The management style was reactive, whereas Ford management was proactive, (some might even say – dictatorial). Merseyside workers do not respect weak management and will always exploit it".

Fred Ward (toolmaker)

Fred joined Vauxhall in 1963 as a machine tool fitter after hanging up his boots as a professional footballer, latterly with West Bromwich Albion. Well-known in football circles, he very soon established the football section of the Vauxhall Sports and Social Club and was its chairman for many years. The football section was probably the most successful of all the club's activities and became semi-professional in later years. They even reached the fourth round of the FA Cup in 2001.

Tony Woodley (production operator, car assembly)

In January 1967, 19 year old Tony started working at Vauxhall Motors' Ellesmere Port factory where his father, George, was the full-time works convenor. Naturally, Tony joined the National Union of Vehicle Builders (NUVB), soon to become part of the Transport and General Workers Union (T&GWU).

Elected onto the branch committee, he became a shop steward, then deputy convenor. By the early 1980s Tony was the T&GWU convenor for the plant, a biannual conference delegate and member of the Vehicle Building and Automotive Group national committee.

Figure 84: Jack Jones (T&GWU General Secretary) visits the plant in March 1977 with Tony Woodley on the extreme right and Ken Boston (factory guide) 2nd from right. On the far left is John Farrell (T&GWU Convener).

As an active member of the Wallasey Labour Party he campaigned to elect its first ever Labour M.P., Angela Eagle. She displaced government minister Linda Chalker in a tough contest which drew on the vital support of trade unionists. The T&GWU chairman persuaded Tony to apply for a job as a full-time union officer based in their Birkenhead office. He was successful. Rising quickly, he became national officer for the Vehicle Building and Automotive Group, and was then promoted to national secretary. This established him as a high profile chief negotiator for the car industry. His no-nonsense attitude has gained him respect and recognition throughout the trade union and labour movements, with a relationship built on mutual respect with Labour cabinet ministers and government departments. Tony's colleagues among the union's leading negotiators showed confidence in him by electing him as shop steward for T&GWU national officers. In 2003 he was elected general secretary of the T&GWU.

In 1980 he was appointed manager and is now chairman of Vauxhall Motors Football Club, a community-based organisation supporting teams from the age of six upwards. An active supporter of the Ellesmere Port Romanian Orphanage Charity, he assists with transportation, delivering much-needed food and dry goods to the physically and mentally disabled children in Romania. A committed and diligent working class representative, Tony has developed into one of the most respected trade unionists in the UK and Ireland.[67]

Bill Thacker (planning engineer)

In February 1962 I completed a five year engineering apprenticeship at Vauxhall Motors Ltd, in Luton and shortly afterwards volunteers were sought to transfer to the coming Ellesmere Port Plant later in the year. I felt that this would be a promising career move and it would give me an opportunity to leave the parental nest. At that time I was a tool test engineer, responsible for the development of cutting tools used in the production of vehicle mechanical components. Shortly afterwards I was asked whether I would take on the responsibility for the development of all the gear cutting tooling at Ellesmere Port. As this seemed a good advancement I leapt at the chance and spent the rest of 1962 under the wing of the current incumbent, Bill Griggs, who was due to retire within a few years and did not relish a family move at that stage.

[67] TW's biography on the T&GWU website www.tgwu.org.uk/Templates/Internal.asp?NodeID=89668

The winter of 1962-3 was one of the longest and coldest for some fifteen years. When I arrived at my lodgings on 30th December 1962 my landlady explained that Chester was between two rivers and that the fall of snow which happened on the morning of my arrival would not stay more than 36 hours – the thaw came in late March! She also explained that Chester landladies had been providing lodgings for camp followers since Roman times. Our mains water supply froze underground after about two weeks of continual sub-zero temperatures. All the household water had to be brought in from a neighbour by bucket so my Saturday bath was taken in the Chester public slipper baths in the old swimming pool building in Union Street, (at 9d. a soak)

I started at the Ellesmere Port plant on the last day of 1962. A few engineers had moved up during the previous weeks and had the temporary use of office accommodation in one of the building contractor's huts. By the time I arrived however, a better temporary office had been built in EA Block. This walled-off area housed all the Vauxhall staff functions for nearly a year until the main office building was functional in 1964. During the summer of 1963 there was a continual fall of dust over all desks and paperwork caused by the groundwork preparations for EC Block. Some of the many new machine tools for the HA Viva were being delivered and installed temporarily in one of the old Belfast hangars until their designated factory location was finished. There was no heating and we had great difficulty getting hydraulic systems to work correctly; snow drifts buried the hydraulic oil reservoirs and kept the viscosity of the oil far too high for any practical use.

Walking through the factory while it was still being built was awe-inspiring. There would be welders in deep trenches working on swarf conveyor systems and coolant tanks for new transfer machines. Others would be dragging four inch diameter high voltage power cables through the overhead roof trusses. One of these workmen was killed when he fell eighteen feet out of the roof and smashed his skull on the corner of a cast iron inspection table. During the winter of 1962-3 some 2,000 new machine tools were delivered, accurately positioned and proved out. The first engine, gearbox and axle units were fully machined and assembled in late April 1963 ready for the start of production of the HA Viva after the 1963 annual holiday closure.

For the first few weeks of 1963 we were able to park our own cars anywhere around the site. This was causing increasing chaos because contractor's vehicles and delivery trucks were unable to gain access to their work areas at crucial times. After just a few weeks the security fence was implemented and employees cars had to use the newly finished car park. There were large cycle racks and bus stops for local services which had commenced running in January. The number of cycle racks was based upon Luton ratio of cycles to employees. However, Ellesmere Port employees came from a far larger catchment area and the cycle racks were never utilised to their full extent. Even after twenty years, by which time employees could have been expected to concentrate in a smaller "travel to work" area, there was always a surfeit of cycle rack spaces. As employee levels increased, to alleviate traffic congestion on North Road, access to the site was also allowed from the Rivacre Road - Hooton Lane junction, but a local farmer always seemed to be driving his cows to the farm for milking at shift starting time, and then again at finishing times, which caused further traffic chaos. The problem was only finally resolved when the M53 opened.

The bus services were soon abandoned by the operators due to lack of passengers. Many of the staff who transferred from Luton moved into the new housing estates at Chester, Waverton, Buckley in Flintshire and on the Wirral. These were up to fifteen miles from the plant. New employees moving house to be nearer the plant generally settled within this fifteen mile radius. This was a much larger area than prevailed at Luton/Dunstable and reflects the greater car ownership of the 1960s.

Summary of the Mini-biographies

Each of these people was successful in their original career fields. Some changed to a completely different activity. This change of career direction was not untypical for Vauxhall Motors and illustrates how people stretched themselves to achieve their full potential, either within the company, in other businesses or in social or trade union activities.

Ellesmere Port and Vauxhall: the Long View

Was the Ellesmere Port Plant Merseyside's lost opportunity?

There used to be a large architectural artist's impression of what the plant might have looked like at some future time. The picture was last seen hanging in a supervisor's office in the press shop. Unfortunately it disappeared in 1997. It showed the three major production buildings which were completed by 1966 and other features which were never built, including extensive factory buildings covering most of the land owned by Vauxhall Motors up to West Road. The picture also showed a rail connection to the nearby British Railways' lines and a berthing quay beside the Manchester Ship Canal. These would have been used for the delivery of cars to other parts of the UK and for export. Vauxhall would then have had excellent motorway, rail and sea connections into one plant. Had the plant been fully developed as envisaged in the painting the labour force could have doubled to 24,000 as had been forecast in the original public inquiry.

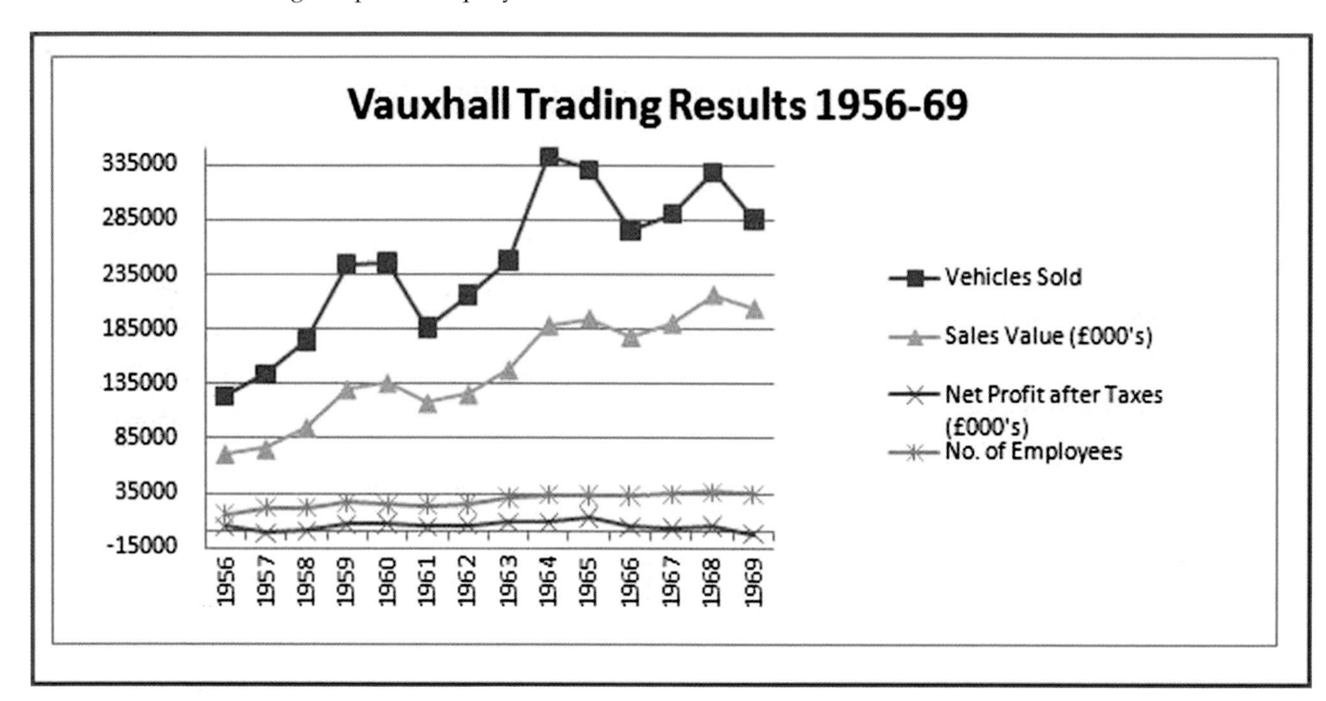

Figure 85: Vauxhall Motors Trading Results 1956-69.

From the graph opposite it can be seen that, while sales, in terms of both units sold and cash received, increased over the years from 1956 to 1964, they fulfilled the company's objectives set out in their 1959 annual report. Then they levelled off from 1964 for the next five years.

Sadly, profits did not increase in line with the sales volumes in future years either and from 1970 to 1986 the company only made a net profit after taxes in one year (1978). During that period the total net loss was a staggering £393m but this was more than recouped by 1990, justifying the difficult decisions taken over the years to persevere with the plant.

During the period 1956-69, the company's payroll increased from 26,251 to 34,156 and peaked at 36,194 in 1968.[68] Such a figure would never to be seen again. By 2009 it had been reduced to about 5,000 actual company employees. This was as a result of closing the Bedford truck operation, ceasing car manufacturing at Luton, ending component manufacture at Ellesmere Port and outsourcing many administration functions. This outsourcing makes it problematic to quantify the number of people actually "working at" Vauxhall because of the jobs which are now carried out off-site by many agencies.

[68] Holden, p208-9

Why wasn't the plant fully developed?

Many factors, both individually and in combination, have influenced decisions that affected the plant's development since 1970:

1. The rise of imported cars, in particular the high specification and quality of Japanese models. Vauxhall lost potential sales and share of an expanding market.

2. Many traditional export markets were lost to newer Japanese manufacturers who were often quicker to react to changing market needs.

3. Disruption to production caused by frequent major labour disputes, often over minor issues.

4. Until the Channel Tunnel opened in 1994, Vauxhall were penalised financially when transporting car parts by cross-channel ferry between plants.

5. The globalisation of car manufacturing, where engines and gearboxes are often manufactured in one country and then built into cars in several others. For example, the Astra cars and vans assembled at Ellesmere Port today, have gear boxes from Japan and engines from Austria. The other side of the coin is that Ellesmere Port now supplies pressings for Astras built in other GM plants across Europe.

6. Vauxhall had previously been a standalone unit, designing and making its own cars. During the 1970s and 80s the company gradually became an integral part of GM Europe. When the design responsibility for cars was transferred to the GM Technical Design Centre at Russelsheim, Germany the company finally lost its autonomy.

7. Labour costs were higher in the UK than in many other countries, particularly in Eastern Europe after the Cold War ended and the Iron Curtain collapsed.

8. It must be concluded however, that the overriding reason for not developing the company and the plan's potential must have been the continuing financial losses.

For these reasons Merseyside was not viewed by GM as a sound location for major investments until 1990 when it was chosen to build the V6 engine. Like all GM facilities, the plant's performance still comes under close scrutiny when it has to compete with other GM plants to win the funding for each new car project. This happens every five to seven years.

Vauxhall has not been alone in having its difficulties. Of the five motor manufacturing companies who expanded into the Industrial Development Areas at the direction of the Board of Trade in the early 1960s, only Vauxhall at Ellesmere Port have weathered the changing misfortunes of the industry. The Rootes Group (later Peugeot Talbot) have closed their Hillman Imp plant at Linwood near Glasgow. Standard Triumph closed their Triumph Spitfire factory at Speke on Merseyside. BMC (Later British Leyland) closed their tractor factory at Bathgate in Scotland. Ford no longer manufactures cars at Halewood on Merseyside, nor anywhere else in the UK. Halewood is now owned by the Indian Tata Group and assembles Jaguar cars and Land Rovers.

If only

...Vauxhall had been able to remain independent from GM's worldwide plans.
...They had managed to continue their excellent labour relations established in the 1940s and 1950s.
...The Japanese had not become smart at designing and building cars in the 1960s.
...The Channel Tunnel had been opened before the 1990s and thus reduced GM's transport costs.

then we may have seen a very much larger plant today. But then history is littered with "ifs"

In September 2009, following GM's chapter 11 bankruptcy, 55% of GM Europe was due to be sold to the Canadian Magna Corporation but GM pulled the plug on the deal at the eleventh hour in November 2009 and decided to retain ownership of GM Europe. GM realised that they did not want to lose the intellectual property, built up over many years, of European small cars to Magna. Magna is partly owned by the Russian Sbersbank (who in turn fund the Russian truck maker GAZ) and GM feared that their technology would be used to design and build competitive cars in Russia which would then be sold in emerging markets in direct competition with GM.

As this book goes to press, in October 2010, the plant looks positively to the future with a leaner and more responsive workforce (just 2,400 direct employees), the sixth generation of Astra in production and an electric car (the Ampera) possibly going to start production in 2011/2.

Figure 86: A publicity photo for the 2012 Vauxhall Ampera.

In April 2010 Vauxhall Motors Ltd. (now renamed GMUK Ltd.) celebrated the 50th anniversary of the announcement that it would come to "The Port". And in the words of Ian Coomber, a former sales director "....it has always been the people of Vauxhall who have had the spirit and determination to see things through. Hopefully, whatever challenges lie ahead, that will never change."

Appendix A

Vauxhall/Bedford model codes:

HA	1963 Viva car and HAV = van variant
HB	1966 Viva car
HC	1971 Viva car
F-type	1957-62 Victor car
PA	1957 Velox and Cresta cars
CA	1952-69 Panel van of 10-12 cwt payload
CF	1969-86 Panel van and chassis cab 10-35 cwt payload
TK	1961-86 Forward control truck of 6-16 tonne GCW

Appendix B

Ellesmere Port Plant building designations:

EA Block	Mechanical components building with an administration office block
EB Block	Boiler house also for the main air compressors and high pressure hot water pumps
EC Block	Mechanical components building including EC basement, EC2 and EC3
ED Block	Main car assembly building, including press and paint shops
EF Block*	Canteen and recreation club office
EG Block*	Office building which mainly housed technical staff

* Both demolished in 2001

Appendix C

Conversion factors:

Linear dimensions
1 foot = 30.5 cm
1000 feet = 305 m

Area dimensions
1 acre = 0.404 hectares
300 acres = 121 hectares

Volumes
1 cubic yard = 0.765 cubic metre
1 hide = 120 acres – a hide is an 11th century measure of area used in Domesday Book. Its size varied in different parts of the country.

Wage level comparison
In 1963 a production worked earned about 7/- per hour and worked a basic 40 hour week earning £14 per week.
In 2009 he earned about £10.90 per hour and worked a basic 36 hour week earning £392.40 per week.

Appendix D

Vauxhall Motors Ltd. Eleven year record – the years of expansion.

Vehicle Sales	1956	1957	1958	1959	1960	1961	1962	1963	1964	1965	1966
Domestic	59,592	59,151	70,713	109,743	117,781	81,113	104,044	135,690	188,342	191,507	175,160
Export	64,051	84,422	103,411	134,912	128,200	105,275	111,930	112,537	154,531	139,476*	100,223
Total	123,643	143,573	174,124	244,655	245,981	186,388	215,974	248,227	342,873	330,983	275,383
Sale Value (£000s)	71,293	76,000	95,070	130,115	135,981	117,903	125,986	147,782	188,328	195,009	178,177
Net profit before taxes	7,086	-2,367	1,121	13,482	14,067	6,888	10,362	10,916	17,927	17,735	3,667
Net profit after taxes	4,634	-1,135	756	6,428	7,159	4,329	5,976	8,267	8,433	13,477	4,435
Gross Fixed Assets	48,490	65,908	65,739	63,655	66,632	76,650	87,215	102,656	108,479	123,908	142,549
Total Payroll (£000s)	12,566	16,654	18,424	22,562	25,005	23,141	25,130	30,123	38,197	39,374	38,520
Employees at year end	16,151	22,084	21,878	26,251	24,573	23,584	24,879	30,843	33,754	33,022	32,859**

*excludes 2184 CKD sets **weekly average during 1966

Bibliography:

Peter J. Aspinal and Daphne M. Hudson, *Ellesmere Port: The Making of an Industrial Borough*, Ellesmere Port Borough Council, 1982

Baileys Magazine for Sports and Pastimes, 1863

Chester Chronicle (both *County* and *Ellesmere Port* editions.), various dates, CRO Ref:- MF204/128, 129 & 132

Ian Coomber, *Vauxhall Bedford Opel Association*, www.vboa.org.uk 2006

Ellesmere Port Pioneer, Microfilm in Ellesmere Port Library, various dates

L. C. Darbyshire, *The Story of Vauxhall 1857-1946*, Vauxhall Motors Ltd., Luton, 1946

Norman Ellison, *The Wirral Peninsular*, Robert Hales & Co., 1955

Goldthorpe J., Lockwood D., Bechhofer F. and Platt J., *The Affluent Worker: Industrial Attitudes and Behaviour*, Cambridge Univ. Press., 1968

B. E. Harris and A. T. Thacker, *Victoria History of the County of Chester Vol 1*, O.U.P., 1987

Richard Hart, *The Vauxhall and Bedford Story*, Farnon Books, 1996

Len Holden, *Vauxhall Motors and the Luton Economy 1900 – 2002*, Boydell Press, 2003

Allison Kelly, *Mrs. Coade's Stone*, The Self Publishing Association, 1990

Liverpool Daily Post, Liverpool public reference library, various dates.

John Peake, *Hooton Park, Early History*, privately pub., 1964, CRO Ref:-CENTRAL 3888

Peter Richardson, *Hooton Park a Thousand Years of History*, Hooton Airword, South Wirral, 1993

The Times, *The Times* online, various dates

Graham Turner, *The Car Makers*, Eyre & Spotswoode, 1963

Hans van Lemmen, *Coade Stone*, Shire Publications, 2006

Vauxhall Motors Ltd., *Annual Reports for 1959-1962*

Vauxhall Motors Ltd., *Vauxhall Facts and Figures – Ellesmere Port*, 1965

Vauxhall Motors Ltd., *The Vauxhall Mirror* and *Management Bulletin*, various dates, 1959-1990

Vauxhall Motors Ltd., *Vauxhall at Ellesmere Port, a guide to the manufacture of Vauxhall cars*, 1977, VM Ref: NE 8/77

Vauxhall Public Affairs Department, *25 Years of Vauxhall at Ellesmere Port*, 1987

Murray Walker, *Unless I'm Very Much Mistaken*, CollinsWillow, 2002

Wikipedia (http://en.wikipedia.org/wiki/hootonpark), 2009

H. E. Young, *Perambulations of the Wirral Hundred*, Henry Young & Sons, 1909

Also many files held by Chester Record Office, Duke Street, Chester

Index